DATA DISSEMINATION IN WIRELESS COMPUTING ENVIRONMENTS

The Kluwer International Series on
ADVANCES IN DATABASE SYSTEMS

Series Editor

Ahmed K. Elmagarmid

Purdue University
West Lafayette, IN 47907

Other books in the Series:

MIDDLEWARE NETWORKS: Concept, Design and Deployment of Internet Infrastructure, *Michah Lerner, George Vanecek, Nino Vidovic, Dad Vrsalovic;* ISBN: 0-7923-7840-7

ADVANCED DATABASE INDEXING, *Yannis Manolopoulos, Yannis Theodoridis, Vassilis J. Tsotras;* ISBN: 0-7923-7716-8

MULTILEVEL SECURE TRANSACTION PROCESSING, *Vijay Atluri, Sushil Jajodia, Binto George* ISBN: 0-7923-7702-8

FUZZY LOGIC IN DATA MODELING, *Guoqing Chen* ISBN: 0-7923-8253-6

INTERCONNECTING HETEROGENEOUS INFORMATION SYSTEMS, *Athman Bouguettaya, Boualem Benatallah, Ahmed Elmagarmid* ISBN: 0-7923-8216-1

FOUNDATIONS OF KNOWLEDGE SYSTEMS: With Applications to Databases and Agents, *Gerd Wagner* ISBN: 0-7923-8212-9

DATABASE RECOVERY, *Vijay Kumar, Sang H. Son* ISBN: 0-7923-8192-0

PARALLEL, OBJECT-ORIENTED, AND ACTIVE KNOWLEDGE BASE SYSTEMS, *Ioannis Vlahavas, Nick Bassiliades* ISBN: 0-7923-8117-3

DATA MANAGEMENT FOR MOBILE COMPUTING, *Evaggelia Pitoura, George Samaras* ISBN: 0-7923-8053-3

MINING VERY LARGE DATABASES WITH PARALLEL PROCESSING, *Alex A. Freitas, Simon H. Lavington* ISBN: 0-7923-8048-7

INDEXING TECHNIQUES FOR ADVANCED DATABASE SYSTEMS, *Elisa Bertino, Beng Chin Ooi, Ron Sacks-Davis, Kian-Lee Tan, Justin Zobel, Boris Shidlovsky, Barbara Catania* ISBN: 0-7923-9985-4

INDEX DATA STRUCTURES IN OBJECT-ORIENTED DATABASES, *Thomas A. Mueck, Martin L. Polaschek* ISBN: 0-7923-9971-4

DATABASE ISSUES IN GEOGRAPHIC INFORMATION SYSTEMS, *Nabil R. Adam, Aryya Gangopadhyay* ISBN: 0-7923-9924-2

VIDEO DATABASE SYSTEMS: Issues, Products, and Applications, *Ahmed K. Elmagarmid, Haitao Jiang, Abdelsalam A. Helal, Anupam Joshi, Magdy Ahmed* ISBN: 0-7923-9872-6

REPLICATION TECHNIQUES IN DISTRIBUTED SYSTEMS, *Abdelsalam A. Helal, Abdelsalam A. Heddaya, Bharat B. Bhargava* ISBN: 0-7923-9800-9

SEARCHING MULTIMEDIA DATABASES BY CONTENT, *Christos Faloutsos* ISBN: 0-7923-9777-0

DATA DISSEMINATION IN WIRELESS COMPUTING ENVIRONMENTS

by

Kian-Lee Tan
Beng Chin Ooi
National University of Singapore

SPRINGER SCIENCE+BUSINESS MEDIA, LLC

Library of Congress Cataloging-in-Publication Data

Tan, Kian-Lee.
Data dissemination in wireless computing environments / by Kian-Lee Tan, Beng Chin Ooi.
p. cm. -- (The Kluwer international series on advances in database systems ; 19)
Includes bibliographical references and index.
ISBN 978-1-4757-7275-3 ISBN 978-0-306-47026-4 (eBook)
DOI 10.1007/978-0-306-47026-4
1. Mobile computing. 2. Mobile communication systems. I. Ooi, Beng Chin, 1961- II. Title. III. Series.

QA76.59 .T36 2000
005.4'376 -- dc21

00-030176

Printed on acid-free paper.

Contents

Preface vii

1. INTRODUCTION 1
   1.1 Reference Architecture of Wireless Computing Environments 1
   1.2 Characteristics of Wireless Computing Environments 3
   1.3 Potential, Benefits and Applications 4
   1.4 Design Issues in Wireless Data Dissemination 7

2. MODELS FOR INFORMATION DISSEMINATION 9
   2.1 Taxonomy of Data Delivery Mechanisms 9
   2.2 Performance Metrics 18

3. DATA BROADCAST SCHEDULING 21
   3.1 Organization of Broadcast for Push-based Broadcast System 21
   3.2 Scheduling Strategies for Pull-based Broadcast System 33

4. INDEXING ON AIR 41
   4.1 Data Organization for Selective Tuning 41
   4.2 Flat Broadcast Programs with Indexes 44
   4.3 Selective Tuning for Skewed Data Access 54
   4.4 Non-flat Broadcast Programs with Indexes 55
   4.5 Selective Tuning Mechanisms for Pull-Based Model 60

5. FAULT-TOLERANT BROADCAST ORGANIZATION 71
   5.1 Fault on Air 71
   5.2 Characteristics of Fault Tolerant Index 73
   5.3 Inter-Index Schemes 74
   5.4 Intra-Index Schemes 76

6. CACHE INVALIDATION MECHANISMS 87
   6.1 A Taxonomy of Cache Invalidation Schemes 88
   6.2 Cache Invalidation Schemes 95

7. BALANCING PUSH AND PULL 115
7.1 Architecture of Integrated Model 116
7.2 The Case for Dynamism 116
7.3 Adaptive Integrated Models 119

8. SUPPORTING RELATIONAL OPERATIONS 131
8.1 Cache Coherency Strategies in Pull-based Systems 131
8.2 Join Processing in Push-based Systems 137

9. CONCLUSIONS 145
9.1 Summary 145
9.2 Research Directions 147

References 153

Index 165

Preface

In our increasingly mobile world, the ability to access information on demand at any time and place can satisfy people's information needs as well as confer on them a competitive advantage. The emergence of battery-operated, low-cost and portable computers such as palmtops and PDAs, coupled with the availability and exploitation of wireless networks, have made possible the potential of ubiquitous computing. Through the wireless networks, portable equipments will become an integrated part of existing distributed computing environments, and mobile users can have access to data stored at information servers located at the static portion of the network even while they are on the move.

Traditionally, information are retrieved following a request-response model. However, this model is no longer adequate in a wireless computing environment. First, the wireless channel is unreliable and insecure. Second, the bandwidth available is very limited varying from 1.2 Kbps for slow paging channels, through 19.2 Kbps (e.g. Cellular Digital Packet Data) to about 10 Mbps for the wireless LAN. Third, the environment is essentially asymmetric with a large number of mobile users accessing a small number of servers. Fourth, battery-operated portable devices can typically operate only for a short time because of the short battery lifespan (before recharging or changing of batteries becomes necessary). Thus, clients may frequently disconnect to conserve energy.

To overcome these limitations, there has been a proliferation of research efforts on designing data delivery mechanisms to support wireless computing more effectively. This book focuses on such mechanisms. The purpose is to provide a thorough and comprehensive review of recent advances on energy-efficient data delivery protocols, efficient wireless channel bandwidth utilization, reliable broadcasting and cache invalidation strategies for clients with long disconnection time. Besides surveying existing methods, the book also compares and evaluates some of the more promising schemes.

The need for such a book is evident. It provides a single source for practitioners, researchers and newcomers on the state-of-the-art in the field. For practitioners, this book explains best practice, guiding selection of appropriate techniques for each application. For researchers, this book provides a foundation for development of new and more effective methods. For newcomers, this book is an overview of the wide range of advanced techniques for information dissemination in wireless computing environments. This book can also be used as a text for an advanced course on information retrieval in mobile and wireless computing, or as a companion text for a variety of courses including courses on distributed systems and mobile and wireless computing.

Organization of the Book

This book consists of nine chapters. Besides the first chapter that sets up the context and the last chapter that concludes, each of the other seven chapters is essentially self-contained and focuses on one aspect of information retrieval. These seven chapters can thus be read and used on their own independently of the others.

In Chapter 1, we provide background on wireless computing in general. We present the reference architecture for wireless computing environments and discuss their characteristics. This chapter also looks at the benefits and promises of wireless computing, and some of the applications that will benefit from wireless computing. It also examines the issues in designing an information retrieval algorithm for wireless computing.

Chapter 2 presents a taxonomy of data delivery mechanisms. The taxonomy is based on our understanding of the design issues and solutions obtained from existing techniques. From the taxonomy, we can obtain a large class of data delivery methods, some of which are known while others have not been explored. We also examine some of the criteria that can be be used to evaluate a data delivery method.

In Chapters 3, 4 and 5, we focus on the organization of data broadcast for information dissemination. There are three main issues that have drawn much research attention, each of which is addressed in respective chapter. The first issue deals with the problem of determining the next object to be picked for dissemination. For the push-based model where the server disseminates data without clients' explicit requests, the emphasis has been on the organization of broadcast data into a broadcast program that minimizes access time. Here, the server generates a broadcast program based on *priori* knowledge of the access probabilities of each object. On the other hand, for the pull-based model where the server disseminates data in response to clients' explicit requests, the server schedules the next object to be broadcast based on some criterion on the outstanding requests. Chapter 3 examines scheduling techniques for both modes.

Chapter 4 looks at the second issue on organizing the broadcast for power conservation so that the effective battery lifespan can be extended. This is important because the effective lifespan of batteries are expected to last two to three hours for most applications. In addition, most current battery research does not predict a substantial change in the available energy in a consumer battery. We shall present techniques that "index" the broadcast data to allow clients to selectively tune to the portion of data that they are interested in. This allows the clients to minimize the time they spent in the active mode and operate in the doze mode most of the time, and hence energy consumption is reduced.

The last issue, presented in Chapter 5, looks at the organization of the broadcast data to deal with noise and errors in wireless transmission. Since the broadcast medium is essentially a sequential medium, missing any data means that the client must wait for the next copy to be disseminated (and incur higher energy consumption). For organization that supports selective tuning, a corrupted bucket along the index path will render the search process invalid, and the clients will have to restart the search from scratch. This is unnecessary especially if the data bucket containing the target objects are uncorrupted. We shall present new data organization techniques that allow the clients to tune to the data quickly (and hence avoid excessively long access time) in the presence of errors in the access path.

In Chapter 6, we focus on cache invalidation schemes for mobile clients who cache frequently accessed data. While caching can reduce the channel contention, the frequent disconnection and mobility of clients complicate the process of maintaining cache consistency: a client who reconnects is no longer certain whether its cache is still valid or not. We shall present a framework on designing cache invalidation schemes, from which we can obtain a large set of cache invalidation schemes. Based on the framework, we shall examine some representative techniques, some of these are energy-efficient solutions.

It is no surprise that neither the push model nor the pull model can perform well in all cases. When the number of clients is large, the push model is the most effective mechanism. On the other hand, when the number of clients is small, the pull model is superior. In Chapter 7, we shall look at how an integrated strategy that balances between the push and pull modes can be adopted. We shall examine the pitfalls of techniques that are *static* (i.e., predetermined amount of bandwidth being allocated to the push and pull models) and provide the case for dynamic strategies (i.e., the amount of bandwidth to be allocated is determined at runtime). We shall also present dynamic mechanisms that are robust to workload and/or access pattern changes.

Chapter 8 is devoted to the processing of relational queries, i.e., queries comprising relational operations (selection, projection, join). Unlike single object retrieval, relational operations typically generate a large number of answer tuples. Moreover, clients queries are expected to exhibit a high degree of data locality, for example canned queries are often performed many times. This facilitates caching of query results. We shall first look at several cache coherency schemes that can be employed to keep the cache content consistent with the relations at the server. We shall also examine several join processing algorithms designed for a broadcast-based system.

Finally, we conclude in Chapter 9 by summarizing the problems and solutions. We also suggest promising research topics that deserve further attention.

KIAN-LEE TAN & BENG CHIN OOI

Chapter 1

INTRODUCTION

We are living in an increasingly mobile world where people travel from place to place. At the same time, information has become such an important asset that many people are craving after to meet their information needs as well as to confer upon themselves a competitive advantage. *Wireless computing* is the phenomenon that enables one to receive and disseminate information even when one moves about — it is mobile computing with untethered communications.

The development of wireless computing systems is made possible by the emergence of lightweight, battery-operated, low-cost and portable devices such as palmtops and PDAs, and the availability and exploitation of wireless networks. Through the wireless networks, portable devices will become an integrated part of existing distributed computing environments, and mobile users can have access to data stored at information servers located at the static portion of the network even while they are on the move.

In this book, we shall focus on mechanisms for information dissemination in wireless computing environments. The purpose is to provide a thorough and comprehensive review of recent advances on data delivery mechanisms in wireless computing environments. In particular, we shall examine techniques that attempt to minimize the *access time* and *energy consumption*, and make efficient use of wireless channel bandwidth. The book also compares and evaluates some of the more promising schemes.

1.1 REFERENCE ARCHITECTURE OF WIRELESS COMPUTING ENVIRONMENTS

Figure 1.1 shows a typical architecture of a wireless computing environment [15]. The wireless environment consists of two distinct sets of entities: a larger number of mobile devices/clients (MC) and relatively fewer, but more powerful, fixed hosts (or *database servers*) (DS). The fixed hosts are connected through a *wired network*, and may also be serving *local* terminals. Some of the fixed hosts, called *mobile support stations* (MSS), are equipped with wireless communication capability. The MSSs and MCs communicate through a wireless communication channel called a *wireless cell* defined over the radio coverage areas of the MSSs. The diameter of a cell, as well as the available bandwidth, may vary according to the specific wireless technology which can either be a cellular connection or a wireless local area network. For example, the diameter of a cell may span a few meters for infrared technology to 1 or 2 miles for radio or satellite networks. On the other hand, the bandwidth of LANs using infrared technology can be of the order of 10 Mbps, whereas WANs have much lower transfer rates [75, 89]. We note that the fixed network is static, whereas the wireless network is mobile since MCs may move from cell to cell.

MCs are expected to be portable devices that are low-cost, lightweight and powered by batteries. MCs can be *dumb terminals* or *walkstations* [57]. The former are diskless hosts (for instance, palmtops) with reduced memory and computing capabilities and has no transmit capability. These MCs will only listen to the channel and select the information requested by users from the data broadcast by the servers. Walkstations, on the contrary, are comparable to classical workstations, and can both receive and send messages on the wireless network. Most of the portable walkstations can also disconnect from the server by operating in a *doze mode* or a *power-off* mode in order to conserve energy.

MSSs can provide both information and commonly used application software for mobile users. They manage and service on-demand requests from mobile clients within the coverage of their cells. Based on the requests, the data/software are retrieved and sent via the wireless channel to the mobile clients. Mobile users can either process the data on their portable devices or tap on the more powerful MSS to perform tasks on their behalf remotely. Each MC is associated with a specific MSS, called *Home MSS*. A Home MSS for a MC maintains specific information about the MC itself, such as the user profile, logic files, access rights, and user private files. Additionally, a user may register as a *visitor* under some other MSSs. Thus, a MSS may also need to keep track of the addresses of users who are currently residing in the cell supervised by the MSS itself.

The wireless channel is logically separated into two subchannels: an *uplink* channel which is used by clients to submit queries to the server via MSS, and a *downlink* channel which is used by MSS to pass the data from MSS to the intended clients. In general, MSS periodically broadcasts frequently demanded information to MCs within its cell who will listen and filter the necessary information. For information that are less frequently accessed, MCs will have to explicitly request for these information.

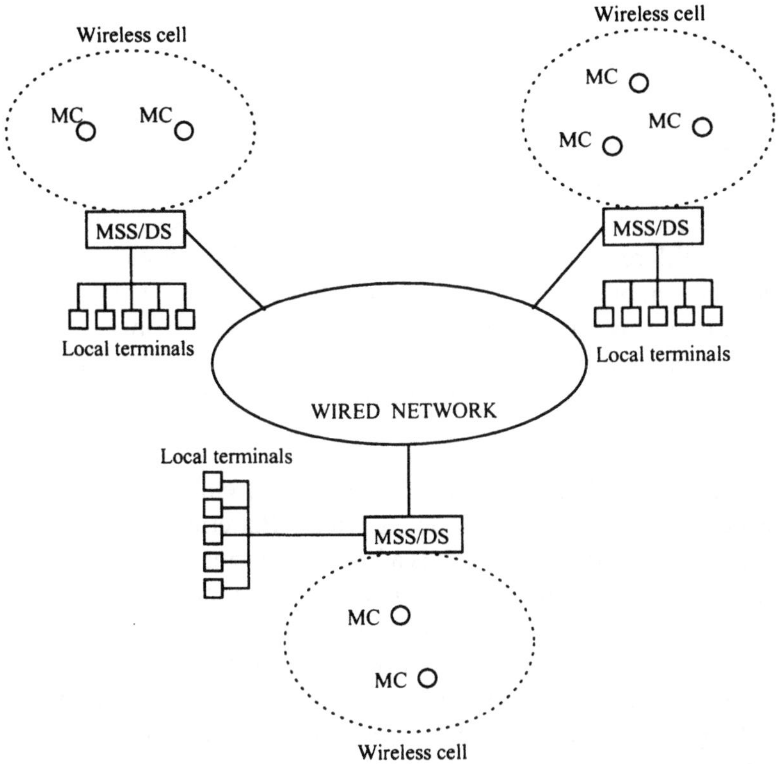

Figure 1.1. Reference architecture of a wireless computing environment.

1.2 CHARACTERISTICS OF WIRELESS COMPUTING ENVIRONMENTS

There are several characteristics that are typical of wireless computing environments. Unfortunately, these are also *limitations* that need to be overcome before the technology can be widely accepted:

- *Limited Wireless Channel Bandwidth.* The bandwidth of wireless channel is very limited. It can vary from ~10 Kbps for wide coverage area (e.g., Mobitex 8 Kbps link or GSM 9.6 Kbps link), through 19.2 Kbps (e.g. Cellular Digital Packet Data) to about 10 Mbps for wireless LAN (e.g., infrared link).

- *Susceptibility of Wireless Communications to Interference.* Wireless transmission is, unfortunately, error-prone. Many factors contribute to this. Signal may be lost due to external interference such as electromagnetic noises or bad weather. Wireless communications are also terrain-sensitive — signals may be affected by tall buildings and trees, hills and valleys. Signals may also be weakened due to the clients' further distance from the server or the high speed at which the clients are moving. Data may also be delayed due to congested network traffics.

- *Asymmetric Communication Environments*. Wireless computing is characterized by its inherent *communications asymmetry*. First, the data volume from servers to clients (downstream direction) is much larger than that from clients to servers (upstream direction). Second, asymmetry arises as a result of the difference in network bandwidth in the downstream and upstream channels — mobile clients are supplied with data at a high rate but can only transmit over low rate channel. Third, there is usually only a small number of servers that provide information to a very large client populations. These servers can easily be "swamped" by an unexpected large number of requests.

- *Limited Effective Battery Lifespan*. Users of wireless computing systems are expected to be carrying portable devices. Batteries are the lifeblood of these devices, and the battery life is affected by virtually all facets of system operations. The complexity and amount of data processed directly affects the energy consumption. The amount and distance of data transmitted also have a direct effect on battery life. As pointed out in [10, 11], under "average user", the power source is expected to last for 2 to 3 hours before replacing or recharging becomes necessary. Unfortunately, as most current battery research does not predict a substantial change in the available energy in a consumer battery, it is important that wireless devices be designed to be energy-constraint. As it is, users will need to carry extra batteries as well as recharging units till battery technology can be improved.

- *Threat to Security*. Since signals travel through the open atmosphere, they can be readily intercepted by anyone. Thus, wireless networks pose more security requirements than its wired counterparts.

- *High Cost*. It is generally more expensive to send data over wireless networks. In some networks (e.g., in cellular phones), network access is charged based on connection time while in others (such as those in radio packets) the rate is based on the number of messages (packets) transmitted.

1.3 POTENTIAL, BENEFITS AND APPLICATIONS

Wireless computing, when fully realized, is expected to change our society into a mobile office that will make information accessible anytime, anywhere and everywhere in the world. Satynarayanan [92] has even envisaged how a businessman's trip may be like in the near future — he can continue his unfinished work, retrieve and record data from the server, access news program, etc, all through his portable computer that he carried with him. While we have yet to reach this goal, most experts agree that wireless computing holds tremendous potential for improving business productivity, services and economic benefits. Some successful examples that exploit mobile and wireless computing technologies include [37]:

- *Financial Industry.* The insurance and mutual fund industries are the first to equip their sales agents with portable devices. For example, Merril Lynch is equipping their agents with Pentium notebooks connected to UNIX servers running Sybase. The sales agent can display and analyze current product and price information right before their customers' eyes. With a modem connection, additional information stored in the central database can be retrieved from client's office or home. This shortens the sales cycle, increases sales productivity and provides quality service to the clients.

  Stockbrokers at the New York Stock Exchange no longer need to clutch pieces of paper and rush to phones to collect order. They can carry handheld PDAs and access information from HP servers through a wireless network.

- *Manufacturing Industry.* Certain business processes can be totally eliminated with mobile computing technology. This will dramatically reduce the time cycles as well as cost. For example, orders can be entered at source and transmitted immediately to shipping and accounting departments. This also has the advantage of minimizing errors and improving the overall quality of information, leading to greater accuracy, fewer complaints, and a reduction in intermediate-tier support staff.

  Manufacturing time cycles can also be cut by using handheld computers and RF-based scanning devices on assembly lines. When used with integrated information systems such as SAP, such practices can result in substantial cost savings, which, in turn, can result in lower prices for consumers and increased competitiveness for producers.

- *Health Care Services.* Devices that monitor patients conditions can be made to transmit critical information directly to nurses and doctors equipped with portable computers or advanced pagers. In this way, doctors and nurses are virtually closer to the patients and can be consulted wherever they may be.

  Wireless technology will also enable instant access to full medical histories of patients through palmtops that will facilitate improved diagnoses. More importantly, critical patient data may need to be transmitted from an ambulance to a hospital. One such system, Remote Patient Monitoring System has been implemented. It allows vital information such as heart rate and blood pressure to be captured either at the scene or en route to the hospital, and transmitted in real time to a hospital emergency room over the Mobitex network. The information is monitored by the consulting doctor, who can make immediate diagnosis and transmit advice back to the ambulance.

Doctors are also increasingly using handheld devices like Newton and pen-based palm pads connected to hospital LANs through wireless PC (PCM-CIA) cards to record patient data.

- *Transportation Industry.* The transport industry can also take full advantage of mobile computing to keep the central databases current. Several taxi fleet owners have successfully deployed computer-aided dispatch for dispatching their cars to customers. The payback is a reduction in dispatch staff required, more accurate dispatching and avoidance of errors in voice communication. With the infrastructure for wireless communications already in place, some taxi services are implementing credit card authorization. Singapore's Comfort Service has introduced a fully automated, satellite-based booking and dispatching system. There are special Comfort Service readers located at pick-up and drop-off points in plazas, airport and other public places. Through these readers, the system identifies the location. Using the satellite-based GPS, the dispatching system locates the nearest available car and sends it the request for a taxi via the Mobitex network. Within seconds, the terminal prints out a slip listing the customer's location and the identification of the assigned vehicle.

  Courier companies such as Federal Express, UPS and DHL as well as trucking industries are also among the early adopters of wireless technology. Parcel tracking as well as vehicle tracking applications based on wireless networks are some of the successful applications that provide customers with superior service. The precise status of the location of vehicles and items carried can provide enquirers with readily available information. Appropriately located freight trucks and courier vans can be informed to pick up items while in transit.

- *Education.* Wireless technology has also been deployed in educational institutes to allow staff and students to learn, interact and access real world resources at any time and any place. The Global Campus project at the National University of Singapore is one such initiative. The wireless technology adopted provides 8 Mbits transmission bandwidth on a share basis, and uses proprietary protocol that is downward compatible for the IEEE 802.11 transmission standard. There are over 30 base stations located at various faculties and offices, each is able to cover an area of 40 m in radius in semi-open office environment or 125 m in radius in open office. Mobile notebook users (students) who are equipped with wireless adaptors (e.g., WaveLAN card) communicate to the campus-wide network wirelessly for lectures that are online via the Web and Video on Demand service, tutorials and assignments at these "strategic" locations.

- *Law Enforcement and Public Safety.* Because of the nature of the work, public safety agencies such as police, fire safety and ambulance services have employed radio network technology for several applications. For example, computer-aided dispatch have been used for police, fire and ambulance vehicles. Queries can also be submitted to local police databases and state/provincial transport systems for vehicle and driver information. In addition, management of records to be stored into the incidence databases can be performed at source.

1.4 DESIGN ISSUES IN WIRELESS DATA DISSEMINATION

There are many challenges that have to be addressed to realize the full potential of wireless computing [11, 56, 57, 58, 67, 83, 91]. These include networking issues (e.g., mobile IP [68, 80], mobility management [8, 22, 63, 64, 73, 86, 97], congestion control in TCP [24]), power management issues (e.g., operating systems level [19, 120], I/O and file systems level [38, 39]), system issues (e.g., disconnection support [70, 104], location dependent queries [55], recovery [1]) and information dissemination issues. We shall focus only on the information dissemination issues, which is the main thrust of this book.

Traditionally, users query database using the *request-response* model. In this model, the clients explicitly request for data objects from the servers, who respond by transmitting the requested objects to the clients. However, because of the unique characteristics (or rather limitations) of wireless computing environments, the basic request-response approach is not the most desirable approach to be deployed in such environments. For example, the request-response approach does not scale well in terms of the number of queries/users that can be supported. Instead, a data delivery mechanism should have the following features:

- *Efficient wireless bandwidth utilization.* The wireless bandwidth can quickly become congested as the client population increases. Moreover, as popular data objects may be continuously requested within a short span of time, multiple copies of the same data objects may be disseminated, leading to poor bandwidth utilization. Mechanisms are needed to effectively utilize the limited bandwidth. One promising approach is to periodically broadcast popular objects without explicit user requests, while those that are less frequently demanded can be accessed using the request-response approach.

- *Efficient and effective scheduling strategies at the server.* The asymmetry between the large client population and the small number of servers can quickly generate huge spikes in the load at the servers. To handle demand-driven requests, the servers must employ efficient scheduling strategies that scale well while balancing the average and worst case response time of queries. Such schemes can exploit the fact that an object that is broadcast will be received by all clients who have requested it to reduce multiple transmission.

- *Energy-efficient data access for battery-powered portable devices.* The need to recharge or change batteries frequently will be a big obstacle to the wide acceptance of wireless technology. Designing energy-efficient solutions at the software level will be critical. Energy efficient solutions make it possible to use smaller and less powerful batteries to run the same set of applications for the same time. Smaller batteries are important from the portability point of view since palmtops can be more compact and weigh less. Moreover, with the same batteries, the unit can run for a very long time without the problem of changing the batteries ever so often. This can result in substantial monetary savings and avoids recharging. In addition, we can minimize the disposal of batteries that can lead to environmental hazard! Software solutions should take advantage of energy-management features that already exist at the hardware level. For example, the algorithms should minimize the time the CPU spent in active mode (and maximize the time the CPU spent in doze mode), keep the receiver off most of the time and spin down the disk, etc.

- *Support for disconnection.* Clients are expected to frequently disconnect, be it *voluntarily* to reduce power consumption or bandwidth use, or *forced* when the portable device enters a region where there is no network coverage. Thus, mechanisms are needed to allow clients to continue to operate even when disconnected. Caching of data is one approach that has been shown to be effective. However, disconnection makes it difficult to ascertain the validity of data objects that are cached when the mobile clients resume connection. Thus, novel cache invalidation strategies are needed.

- *Support for secured and reliable transmission.* While existing techniques for secured transmission on wired environment can be applied to wireless context, the fact that the transmitted information can be picked up by everybody calls for more secure measures to be developed. It is also necessary to deal with the unreliable network in order to minimize retransmission and long waiting time. The basic approach of injecting redundancy into the broadcast has to be adapted and controlled since replicating broadcast information implies longer broadcast length.

From the above discussions, it is clear that we need novel data delivery mechanisms in wireless environments. This book is devoted to dealing with most of these issues and to review the various approaches that have been adopted in the literature.

Chapter 2

MODELS FOR INFORMATION DISSEMINATION

Traditionally, the mode of information dissemination has largely been one of *request-response* style, i.e., users submit queries to servers that respond with the answers. However, in recent years, different models of data delivery have been explored, particularly the *periodic push* model where servers repetitively disseminate information to clients without any explicit requests. Interestingly, these two models are just two extremes among many data delivery methods that can be adopted to disseminate information.

In this chapter, we shall first present a taxonomy of data delivery mechanisms. We shall then discuss the performance metrics that have been/can be used to evaluate the effectiveness of these mechanisms in wireless computing environments.

2.1 TAXONOMY OF DATA DELIVERY MECHANISMS

Franklin and Zdonik [46, 47] have presented a taxonomy of data delivery mechanisms. However, it has not adequately captured the full design space of data delivery mechanisms. We shall present an alternative taxonomy based on the design issues that need to be addressed in developing a data delivery mechanism. These design issues are derived based on our understanding of the problem and the schemes that have been proposed in the literature. We have identified four main issues that can be broadly categorized as *data-oriented*, *mechanics-oriented*, *organization-oriented* and *bandwidth-oriented* issues. From the solutions to these issues, we can obtain a large collection of data delivery mechanisms, some of which are known and widely used, while others have yet to be explored. We believe such a framework can help to identify the similarities and differences between different methods, and provide necessary foundation or building blocks for designing new mechanisms. Figure 2.1 summarizes the issues and possible solutions to these issues.

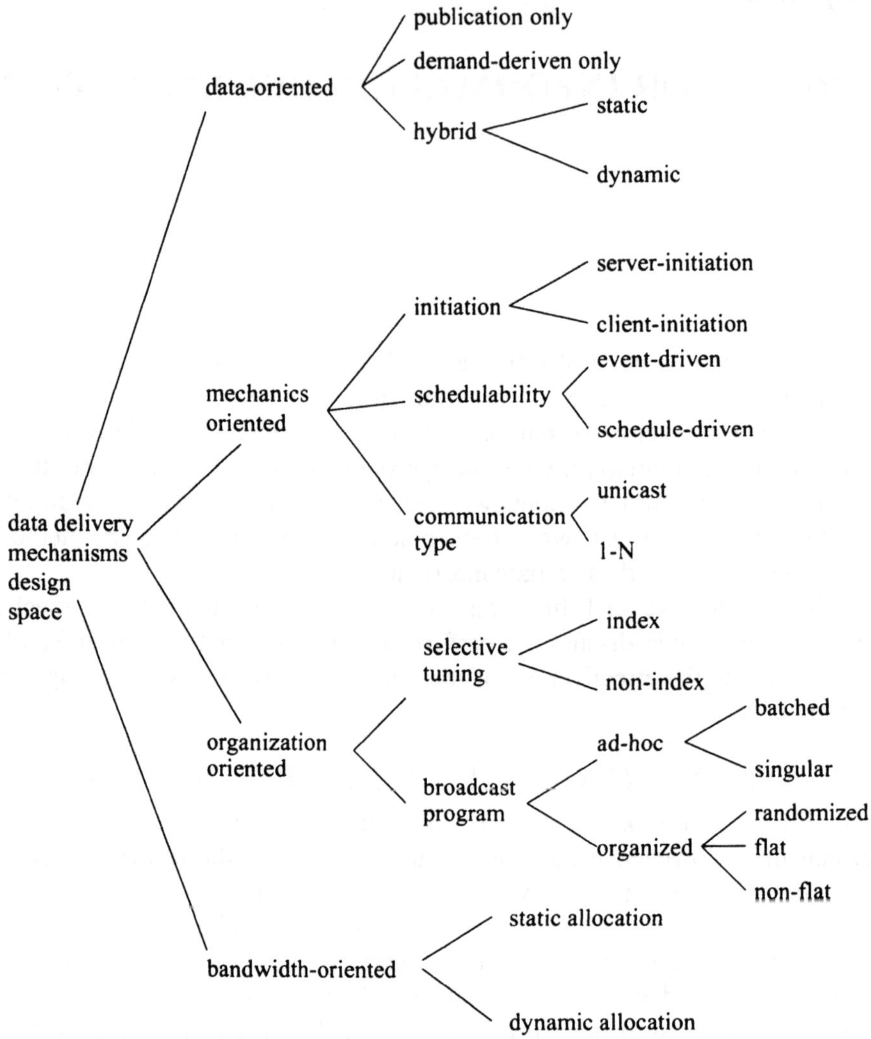

Figure 2.1. A taxonomy of data delivery strategies.

2.1.1 Data-Oriented Issues

Data objects are essentially partitioned into two groups [59]: *publication group* and *demand-driven group*. Objects in the publication group are broadcast by the server regardless of whether there are requests for them or not. To access these objects, clients listen to the downlink channel and filter the incoming data stream to obtain the desired objects. On the other hand, objects in the demand-driven group are provided to clients upon requests, i.e., clients must submit queries explicitly to servers for the objects they want, if those objects belong to this category.

We can thus derive three possible ways in which data can be grouped:

- *Publication-only.* All the data objects are placed in the publication group, and objects are published without explicit requests. This approach has been adopted in [52, 60, 61, 111, 114, 119].
- *Demand-driven-only.* All the data objects are placed in the demand-driven group, and objects are only broadcast in response to explicit requests. Examples of this approach can be found in [15, 26, 123, 109].
- *Hybrid.* In a hybrid approach, some objects are allocated for publication, while the remaining objects are to be requested before they are delivered. [5, 27, 34, 101] are examples of works that have employed this technique. As we shall see shortly, both the publication-only and demand-driven-only approaches are special cases of the hybrid approach. However, because of the large number of works that belong to these categories, it simplifies our presentation by treating them separately.

The hybrid approach can be further divided into two categories:

- *Static.* In the static approach, the number and the set of objects are fixed, and there is no mechanism to vary either the number or the set of objects at runtime. For example, we can broadcast the entire file periodically (which corresponds to the publication-only approach), or we can provide data solely based on demand basis (i.e., the demand-driven-only approach). In [5, 59], the hybrid approach predetermines a fixed number of objects in the publication group based on the average access patterns of requests.
- *Dynamic.* We can also adopt a very dynamic approach where the number of objects and the set of objects to be included in the publication group are determined at runtime. This can be based on the number of clients and/or access patterns. For example, if more clients are tuning in, then it may be more effective to broadcast more objects. On the contrary, if there are very few clients, providing objects on demand can better utilize the channel bandwidth. This approach has been explored in [27, 100, 101, 113].

The tradeoffs between the two hybrid approaches are obvious. The static approaches are simple but rigid. They are unable to adapt to changing workload or access patterns, and are unlikely to scale well. Moreover, it is difficult to determine the optimal number and set of objects to be placed in the publication group: publishing too many objects on air may waste the bandwidth as some of them may be less or even not demanded; on the contrary, broadcasting too few objects may lead to poor utilization of bandwidth as multiple copies of popular objects in the demand-driven group may be broadcast repeatedly. These techniques are, therefore, suitable for stable environments. However, because mobile clients move from cell to cell, the access patterns within a cell is likely to be highly varied, making this approach unattractive for wireless computing.

The dynamic approach is expected to be more robust to the changing environment. However, it requires complicated runtime algorithms to monitor the workload/access patterns and react to these changes effectively and efficiently. We note that the workload/access patterns for the demand-driven group can be monitored, but this cannot be done for the publication group (unless users submit explicit feedback to the server on their "hits" and "misses" in accessing data from the publication group).

2.1.2 Mechanics-Oriented Issues

This dimension deals with the mechanics for disseminating information, and considers three issues. These issues are also those that Franklin and Zdonik include in their design space [46, 47]. These issues are:

- *Delivery Initiation.* Information dissemination can be initiated by the client (*client-initiation*) or by the server (*server-initiation*). Under the client-initiation approach, clients "pull" the desired data objects from the server by submitting queries to the server, the server accesses the relevant objects from the database and returns them to the clients. The server-initiation approach operates in a different manner: the server initiates data dissemination and sends data objects to clients possibly without any explicit requests from clients. To receive the objects, clients must listen to the broadcast channel to filter the incoming stream for their desired data. The client-initiation approach is effective when the client population is small; otherwise, the communication channel and server can become a bottleneck quickly. Moreover, it requires an uplink channel to be available and that the client must have transmission capability. In addition, clients need to be able to know what they want to retrieve. On the contrary, the server-initiation approach is an effective mechanism for large client population and avoids the limitations of the client-initiation scheme. However, it is also limited by three problems. First, sending irrelevant data results in poor use of the channel bandwidth. Second, data may not reach the clients in time (since it depends on when the data are pushed). Thus, the server needs to be able to predict accurately the needs of clients. Third, there is no way for the server to know how the data that are pushed out are being accessed. The server-initiation approach has been explored in [60, 61, 114, 115, 119], while the client-initiation approach is adopted in [15, 26, 123, 109]. While the server-initiation mechanisms are commonly used for disseminating data in the publication group, there are works where clients may initiate dissemination from the publication group [5].

- *Schedulability.* This dimension examines the schedulability of the data, i.e., whether the objects are *event-driven* or *schedule-driven*. In the former, there is no predetermined schedule on how objects are to be disseminated. Objects are disseminated in response to events such as requests or triggered by updates. The latter delivers data based on some predetermined schedule. For example, information may be sent out daily or weekly or information may be polled periodically as in remote-sensing applications.

- *Communication type.* The last issue concerns the type of communication, whether it is 1-1 (*unicast*) or 1-N communication. With unicast, data is sent from the server to a client, while 1-N communication allows the same data to be received by multiple clients. 1-N communication comes in two flavors. In *multicast* mode, the multiple recipients are known in advance and the data is sent only to these specific clients. With *broadcast* mode, information is sent to all clients (an unknown and possibly unbounded number) listening to the channel. Obviously, 1-N communication is useful when multiple clients share the same interests; otherwise, the clients will be continuously being interrupted by unwanted data.

2.1.3 Organization-Oriented Issues

The objects to be disseminated have to be organized for dissemination. There are two factors to consider:

- *Broadcast program.* The queried objects to be broadcast may be organized either to minimize bandwidth utilization or improve performance (access time). This can be done in an *ad-hoc* manner either by simply sending the answers of each query to the clients or *batching* the answers of multiple queries to minimize repeated transmission of the same object. The former *singular* approach has been widely adopted (e.g., in [15, 26, 123]), while the latter has been considered in [6, 7, 109].

  Alternatively, the objects can be organized based on some program/schedule. A program is *cyclic* if the objects disseminated follows a regular, repeating schedule, while an *acyclic* program repeatedly disseminate objects without a predetermined schedule. Cyclic programs may also be *flat* or *non-flat*. In flat programs, all objects are of equal importance and disseminated once in a *broadcast cycle*. On the contrary, a non-flat program favors objects with higher access frequencies. Hence, in a broadcast cycle of a non-flat broadcast, while all objects are broadcast, some will appear more often than others. Acyclic programs are typically non-flat in nature. For simplicity, we shall refer to acyclic programs as *randomized* (or repetitive) programs. For cyclic programs, because a data object appears repeatedly in the broadcast, a client that misses the object may be able to get it in the same cycle (for non-flat programs) or at most in the next cycle. The regularity of a cyclic program also makes it easier to design mechanisms that allow clients to selectively tune to the desired portion of the data (rather than examining the entire program). Non-flat programs yield shorter *access time* (i.e., the time the client initiates the query till the desired object is downloaded) for popular objects as compared to flat programs at the expense of longer access time for objects that are less frequently accessed. Non-flat programs also provide a better average access time than flat programs. However, it has higher bandwidth requirement as its broadcast cycle length is longer than that of flat programs. Flat programs have been studied in [61, 60] while non-flat programs have been explored in [2, 111, 115]. Randomized programs generated using probabilistic approaches can be found in [59, 121] and those constructed by scheduling algorithms appeared in [52, 119].

- *Selective Tuning.* In a wireless environment, a client listening to the channel need to examine every object that is disseminated. This process requires the CPU to be in the full operation (*active*) mode, which is a very power-consuming operation. Since the client is typically interested in only a small subset of the disseminated data, the overhead of scanning the other objects is wasted. To minimize the scarce energy resources in portable devices, methods that index data "on air" have been proposed to organize the disseminated data so that clients only need to selectively tune to the desired data [26, 60, 61, 77, 111, 115]. This is realized by multiplexing meta-data with the data objects to be broadcast. While selective tuning mechanisms minimize energy consumption, the extra meta-data imposes additional requirement on the channel bandwidth.

  In [60], the indexing schemes are based on tree structures, while those in [61] are hash-based. A highly parameterized scheme, *flexible indexing scheme*, that facilitates variability of the organization of the broadcast to control energy consumption and access time is also proposed in [61]. The work in [77] employs *signatures* as an indexing tool instead. While all these works looked at flat programs, [111, 126] adapted schemes in [60, 61] for non-flat programs. Index schemes for unclustered indexes have been proposed in [62].

2.1.4 Bandwidth-Oriented Issues

To support two groups of data, i.e., publication group and demand-driven group, the downlink bandwidth must be distributed between the two groups. Let the bandwidth assigned for the downlink channel, the publication group and the demand driven group be β, β_p and β_o respectively. Note that $\beta = \beta_p + \beta_o$. The downlink channel can be viewed as time slots, and objects are broadcast in slots such that for every $\lceil k \cdot \beta_p/\beta \rceil$ slots allocated to broadcast objects in the publication group, $\lceil k \cdot \beta_o/\beta \rceil$ slots will be reserved for the demand-driven group, for some integer k. There are two possible methods to determine the bandwidth allocations:

- *Static allocation.* In this category, the bandwidth is assigned to the two groups in advance, and there is no mechanism to change the allocation during runtime. Such an approach is simple, but may lead to poor bandwidth utilization as access patterns and workload change. The schemes in [5, 59] adopt such an allocation.

- *Dynamic allocation.* In this category, the bandwidth is assigned to the two groups based on some optimization criterion. In this way, the workload and access patterns can be factored into the decision. For example, if the workload is high, then more bandwidth can be allocated to the publication group. Different allocation techniques can be adopted. One approach is to allocate the bandwidth proportionally based on characteristics of accesses on the two groups. Proportionality comes in different flavors: the bandwidth can be assigned proportionally based on the *number of clients*, *access patterns*, *number of objects in each group* and a combinations of these. As another example, bandwidth can also be allocated such that the average access time for the two groups are the same. Dynamic hybrid approaches [34, 101] employ this allocation scheme.

2.1.5 Data Delivery Strategies

From the taxonomy, we can derive a large number of data delivery strategies from the product of the number of alternatives in each dimension. Many of these strategies are known methods, while some of them have not been investigated. For example, there is no reported work on hybrid data-oriented schemes that exploit selective tuning for both groups of data. We also note that some of the combinations are not feasible. For example, it does not make sense to design publication-only schemes with dynamic bandwidth allocation schemes.

We shall look at how some of the existing data delivery mechanisms can be classified using the framework. These are summarized in Tables 2.1—2.3. Table 2.1 summarizes mechanisms whose data-oriented dimension is the publication-only approach. Thus, for the bandwidth-oriented dimension, all the bandwidth are statically allocated to the publication group. As such, we ignore these two dimensions in the table. Moreover, since these techniques are all server-initiation methods, we shall refer to them as push-based methods. We note that the first three mechanisms have also been classified in [46, 47]:

- *Publish/subscribe.* In the publish/subscribe model of information dissemination, users provide information in the form of a profile that represent his information needs. As new data are continuously received by the server, it filters them against the user profile and delivers those matching the users needs to them. The server initiates data dissemination. It is considered event-driven since information is only sent whenever there are data matching the users' profile and this cannot be predetermined. The approach is 1-N connection oriented, but some implementations (especially on the Internet) employ unicast messages to disseminate information to multiple clients (e.g, Internet email listing). There is no selective tuning employed and the objects are transmitted in singular ad-hoc manner.

- *Email listing digest.* Some systems allow a list manager to set up a schedule for sending "digests" on a regular basis to a mailing list. In these systems, digests are pushed (server-initiation) to the individual clients (unicast) in the mailing list regularly (schedule-driven). There is no support for indexing, and data are transmitted in singular ad-hoc manner.

- *Broadcast disks.* Broadcast disks refers to the class of server-initiated, schedule-driven mechanisms that publish information to multiple clients. The basic approach does not support indexing. Data, however, are organized into a cyclic non-flat schedule.

We note that none of these three data delivery methods employs selective tuning. The algorithms in [52, 119] generate non-flat/randomized broadcast programs and there is no support for selective tuning. While the methods in [30, 60, 61, 77] employ selective tuning on flat broadcast programs, the schemes proposed in [111, 126] exploit both non-flat broadcast programs and selective tuning.

Table 2.2 summarizes mechanisms whose data-oriented dimension is the demand-driven-only approach. In this case, for the bandwidth-oriented dimension, all the bandwidth are statically allocated to the demand-driven group. As before, we ignore these two dimensions in the table. Moreover, since these techniques are all client-initiation methods, we shall refer to them as pull-based methods. Again, the first four mechanisms have also been classified in [46, 47]:

- *Request/response.* The traditional request/response mechanisms essentially ignore the publication group, i.e., all data objects are demand-driven. Data is disseminated by client's request (client-initiation) whenever the clients need the objects (i.e., event-driven). Typically, a client will pull its own objects (i.e., unicast connection). The request/response method also does not facilitate selective tuning, and objects for each request are disseminated in an ad-hoc manner (as and when there is a request). By using a 1-N connection with the request/response scheme, clients who have not made explicit requests on the objects can "snoop" on the answers from other requests. This gives rise to the *request/response with snooping* strategy.

Table 2.1. Push-based data delivery mechanisms.

Method	*Mechanics Oriented*	*Organization Oriented*
publish/ subscribe	server-initiation, event-driven, both unicast and 1-N	non-index, singular ad-hoc
email listing digest	server-initiation, schedule-driven, unicast	non-index, singular ad-hoc
broadcast disks [2]	server-initiation, schedule-driven, 1-N	non-index, non-flat
[52, 119]	server-initiation, schedule-driven, 1-N	non-index, non-flat/randomized
[30, 61, 60, 77]	server-initiation, schedule-driven, 1-N	index, flat
[111, 126]	server-initiation, schedule-driven, 1-N	index, non-flat

- *Polling.* The polling mechanism is commonly found in applications that periodically send requests to servers to obtain status information or to detect changes in data values. These applications include remote sensing or control applications. It is similar to the response/request model in all dimensions except the schedulability issue under the mechanics-oriented dimension. In polling, clients typically follow a schedule in submitting requests, i.e., it is schedule-driven. While the basic polling mechanism is unicast, 1-N connection will allow other clients to snoop to the broadcast information.

We note that only techniques studied in [112] adopt selective tuning. Methods proposed in [6, 7, 109] incorporate batching techniques in organizing the data to be disseminated. Further, the schemes in [26] employ selective tuning techniques on the invalidation reports that are used to validate client cache content, but not in the dissemination of the objects.

Finally, Table 2.3 summarizes mechanisms that support more than one mode of information dissemination, which we shall refer to as *integrated* methods. We note that, with the exception of [59], all the other schemes have some form of runtime mechanisms to adapt to workload and/or changing access patterns. In particular, the approaches in [113] are robust to only workload changes,

Table 2.2. Pull-based data delivery mechanisms.

Method	*Mechanics-oriented*	*Organization-oriented*
request/ response	client-initiation, event-driven, both unicast and 1-N (w/snooping)	non-index, singular ad-hoc
polling	client-initiation, schedule-driven, both unicast and 1-N (w/snooping)	non-index, singular ad-hoc
[15, 123]	client-initiation, event-driven, unicast	non-index, singular ad-hoc
[26]	client-initiation, event-driven, unicast	(partial) index, singular ad-hoc
[6, 7, 109]	client-initiation, event-driven, unicast	non-index, batched ad-hoc
[112]	client-initiation, event-driven, unicast	index, singular ad-hoc

while those in [27, 34, 101] can also adapt to access pattern changes. It is interesting to note that most of the schemes disseminate published data using randomized broadcast programs. Only the schemes in [27, 107] employ non-flat broadcast programs. In addition, only a few schemes adopt selective tuning [27, 34, 101, 107]. We also note that all schemes, except [5], partition the database into two groups. Even though [5] publishes all data, it provides an uplink channel for users to submit requests if their objects are expected to take a long time to arrive (e.g., when they have been missed). By allocating some channel bandwidth for these requests, the server can respond to these requests before their objects are scheduled for broadcasting.

2.2 PERFORMANCE METRICS

In wireless computing environments, several criteria can be used to evaluate the performance of a data delivery method:

- *Responsiveness.* The success of a data delivery scheme is determined by its ability to get the requested data to the clients quickly. In this regard,

Table 2.3. Integrated data delivery mechanisms.

Method	*Data Oriented*	*Mechanics Oriented*	*Dissemination Oriented*	*Bandwidth Oriented*
Adaptive scheme [59]	static	Publication server-initiation & schedule-driven, 1-N Demand-driven client-initiation & event-driven, 1-N	Publication non-index, randomized Demand-driven non-index, singular ad-hoc	static
IPP [5]	static	Publication server-initiation & schedule-driven, client-initiation & event-driven, 1-N	non-index, randomized	dynamic
[27, 113]	dynamic	Publication server-initiation & schedule-driven, 1-N Demand-driven client-initiation & event-driven, unicast	Publication index, non-flat Demand-driven non-index, singular ad-hoc	dynamic
[34]	dynamic	Publication server-initiation & schedule-driven, 1-N Demand-driven client-initiation & event-driven, unicast	Publication non-index, randomized Demand-driven non-index, singular ad-hoc	dynamic
[101]	dynamic	Publication server-initiation & schedule-driven, 1-N Demand-driven client-initiation & event-driven, unicast	Publication index, randomized Demand-driven non-index, singular ad-hoc	dynamic

the first important metric is the *average access time*. The average access time is the amount of time spent, on average, from the instant the request is made to the time that the requested object is received. The second metric is the *worst-case access time* that measures the maximum access time for any client request to be satisfied. The former provides a metric for the *overall system* performance, while the latter indicates *individual performance*.

- *Robustness*. The effectiveness of a scheme is also determined by how well it adapts to workload or environmental changes. The scheme should be able to support increasingly larger population (i.e., scalable), or be robust to changing access patterns. A scheme that is rigid will be useful only in *static* environments where the workload is fairly stable, while a robust scheme is necessary for *dynamically* changing environments.

- *Packet Efficiency*. In cellular telephony, one of the most important criteria is the cell capacity that measures the number of telephone calls which can be handled by the cell per unit of time. In wireless computing, a query will be the analog of a telephone call. Thus, the criterion *queries per Hz* that measures the cell capacity in terms of the number of queries that the local mobile support station can handle per unit of time is expected to provide better indication of the packet efficiency.

- *Power Efficiency*. Battery power, being a precious resource for portable devices, has to be minimized. As such, the metric *queries per watt* has been proposed [59] to measure the power efficiency of a method. Queries per watt essentially indicates the number of queries that a client can process from 1 Watt battery.

- *Tuning Time*. Another metric that is commonly used as an indication of the energy consumption of an information dissemination strategy is the *tuning time*. The tuning time measures the amount of time that a mobile client listens to the channel. Thus, it measures the time during which the client stays in the active mode and therefore determines the power consumed by the client to retrieve the relevant data.

Ideally, we hope to have a robust and energy efficient data delivery method that minimizes both the average and worst-case access time. However, some of these metrics may conflict. For example, building "indexes on air" for broadcast data can help to minimize energy consumption but the extra meta-data lengthens the broadcast length and hence increases the access time. Thus, it is a challenge to design schemes that balance these factors.

Chapter 3

DATA BROADCAST SCHEDULING

The wireless computing environment is essentially a broadcast data delivery environment. There are two basic modes in which information can delivered to mobile clients. In the first mode, the *push* mode, a server continuously pushes information in a broadcast channel and the clients access it by tuning in and waiting until the information they are interested in is transmitted. The server follows a *schedule* (or *program*) that attempts to match the user access statistics in order to reduce the access latency. The program determines the order in which objects should be broadcast. The main challenge is to design scheduling algorithms that can construct the optimal program that minimizes the average access time.

In the second mode, the *pull* mode, the server delivers information to mobile clients in response to their submitted requests. From the (outstanding) requests, the server determines the next object to be picked for dissemination based on some criterion. The challenge in this case is to design scheduling algorithms that determine the object which can greatly improve the effectiveness of the system when it is picked for broadcasting.

This chapter looks at the problem of scheduling the data broadcast for both the push and pull modes of information dissemination. In Section 3.1, we examine how data should be organized to minimize the access time under the push model. In Section 3.2, we present scheduling techniques for the pull model.

3.1 ORGANIZATION OF BROADCAST FOR PUSH-BASED BROADCAST SYSTEM

Disseminating information using the push model is not something new. The newspapers that we read, the programs that we listen to over the radio and watch over the television employ the push technology. In computing, we have witnessed its application in Bellcore's Datacycle database machine [23, 53], the Videotex and Teletex systems for distributing data over broadcast media [12, 121, 122] and the Boston Community Information System at MIT [48]. More recently, the push technology has also been deployed on the Internet and Web (e.g., PointCast's webcasting [84]) to display headlines, weather forecasts, stock prices, etc. on idle desktop machines turning them into active information "kiosks".

To exploit the larger downstream bandwidth capacity to *continuously* and *repeatedly* broadcast data to clients, the server constructs a *broadcast program* for the data objects. The broadcast program essentially determines the order in which objects should be broadcast. Under the push model, the server is not aware of the user's requests, and so the broadcast program is designed based solely on the distribution of the object requests. For the rest of this section, we adopt the following notations. There are N data objects and the ith object has size l_i. Moreover, it takes one time unit to broadcast an item of unit length. In addition, the broadcast consists of *broadcast cycle* of size L time units. We note that it is possible for an object to appear more than once in a broadcast cycle. Each appearance will be referred to as a *copy* of the object. The number of copies of an object i in a broadcast cycle is called its *frequency* and is denoted as f_i. The size of a broadcast cycle is therefore given by $M = \sum_{i=1}^{N} f_i l_i$. For each request, the probability that object i is requested is assumed to be q_i. We assume that $q_1 \geq q_2 \cdots \geq q_N$, and $f_1 \geq f_2 \cdots \geq f_N$. Finally, the *spacing* between two copies of an object is the time it takes to broadcast information from the beginning of the first copy to the beginning of the second copy. We denote the spacing between the jth copy of an object i and the next copy of i as s_{ij}. Figure 3.1 shows an example of a part of a broadcast program.

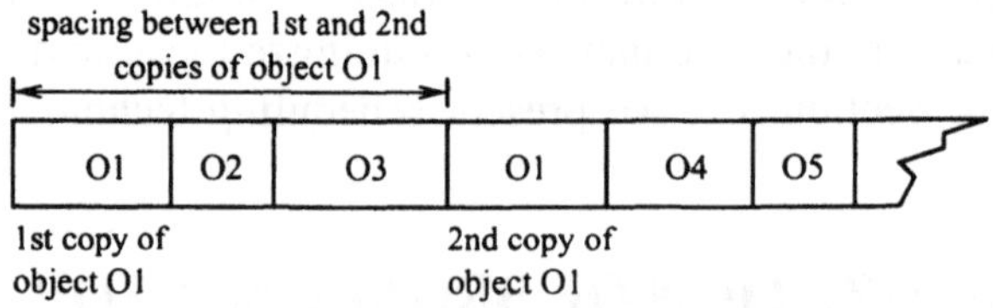

Figure 3.1. Part of a broadcast cycle.

3.1.1 Properties of Broadcast Program

In the simplest scenario, the server can send the union of all objects that may be needed by its clients in a periodic fashion. Such a program has been termed the *flat* broadcast program. With a flat broadcast, the average access time for an object is the same for all objects and is equal to half the broadcast period, while its worst-case access time is equal to the broadcast period. However, the flat organization makes the process of adding new data objects to the broadcast program very simple: simply insert the objects into the broadcast program and lengthen the broadcast period. Most of the early works on broadcast-based systems have adopted this flat broadcast approach [23, 53, 62].

In practice, very often, some objects are more frequently accessed than others. For example, we can expect the 80%-20% rule to hold, i.e., we have 80% of the users interested in only 20% of the objects. In this case, it makes sense to broadcast the frequently accessed objects more regularly than those that are less popular. This *non-flat* broadcast program will reduce the average/worst-case access time of popular objects as compared to the flat broadcast program, but the expected delay in accessing less popular objects is longer than with a flat broadcast.

Conceptually, the generation of non-flat broadcast programs can be addressed as a *bandwidth allocation problem*: given the client access probabilities, the server determines the optimal percentage of bandwidth that should be allocated to each object. The broadcast program can then be generated according to those bandwidth allocations. The challenge is to find the optimal program that minimizes the expected access time. Figure 3.2 shows three sample broadcast programs for a data set containing three equal-sized objects. Program (a) is a flat broadcast, program (b) and (c) are both non-flat broadcast programs that broadcast object A twice as often as objects B and C. Program (b) is a *skewed* broadcast, in which all copies of object A are clustered together in the broadcast. In contrast, program (c) is *regular* where there is no variance in the inter-arrival times for each object. In [3], several interesting observations were made. First, as the access probabilities become increasingly skewed, the non-flat programs perform increasingly better. Second, the regular programs always outperform the skewed programs. Generating non-flat programs has been studied in [2, 118, 119, 121].

We note that it may not be practical to construct non-flat broadcast programs for large systems with huge number of objects. In such systems, the (near) optimal solution can be a very complicated broadcast program. In addition, the process of adding new objects may require a complete overhaul of the broadcast program.

An alternative approach to generating non-flat broadcast programs is to coalesce objects with similar access probabilities into one partition. Objects in the same partition will be broadcast in the same frequencies (in a broadcast cycle) and hence share the same average/worst-case latencies. The resultant program obtained, referred to as the *non-flat coalesced* broadcast program, is no longer optimal. However, the complexity of generating such a coalesced program is significantly reduced. Moreover, adding a new object is much simpler than the non-flat broadcast program since the object is inserted into the appropriate partition. This is the approach adopted by *Broadcast Disks* [2, 3].

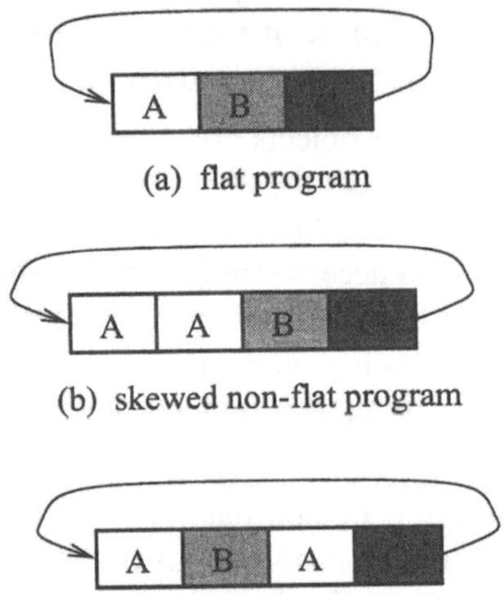

(a) flat program

(b) skewed non-flat program

(c) regular non-flat program

Figure 3.2. Broadcast programs.

3.1.2 Generation of Broadcast Programs

The simplest approach to broadcasting data objects is to probabilistically pick an object for transmission [59, 121]. This algorithm operates as follows. At each broadcast instance, object i is selected with probability p_i, where p_i is determined based on the access probability. However, this method results in acyclic broadcast. It may also lead to arbitrarily large access time. Moreover, it has been shown to be inferior to other more systematic approaches [121]. For pedagogical reasons, we shall use the following simple running example throughout this section:

EXAMPLE 1 *Let a data file comprises 33 objects (as shown in Figure 3.3) with keys that range from 11 to 91. We assume that all objects are of the same size. Here, we have logically split the objects into 3 partitions, A (with 3 objects), B (with 6 objects) and C (with the remaining 24 objects). We further assume that objects in a partition have the same access frequencies. Objects in partitions A, B and C have access frequencies of 4/21, 1/21 and 1/168 respectively.* □

Generating non-flat broadcast programs

Wong's algorithm [121]. In [121], Wong showed that, for fixed sized objects, the access time can be minimized if

A

14
29
69

B

20	27
22	31
25	33

C

11	35	45	50	60	65	75	85
17	38	47	52	62	67	80	88
19	41	48	55	63	72	82	91

Figure 3.3. A sample set of objects to be organized for broadcast.

- copies of an object are equally spaced. In other words, given a broadcast cycle length of L, and object i has frequency f_i, its copies are placed at every L/f_i slot in the cycle. Note that this is not always possible when $i > 1$ because one or more positions needed by object i may have been assigned to objects with lower indices already.

- for any two objects i and j, $f_i/f_j = \sqrt{q_i}/\sqrt{q_j}$. This expression provides a means to determine the number of copies of an object to disseminate in a broadcast cycle.

Figure 3.4 shows Wong's algorithm that generates a cyclic broadcast of length L given the access probabilities, $q_1, \ldots, q_N$.

Input: cycle length, L; access probabilities, $q_1, \ldots, q_N$
Output: cyclic non-flat program

1. select integer f_i such that $\sum_{i=1}^{N} f_i = L$
   and f_i/f_j as close to $\sqrt{q_i}/\sqrt{q_j}$ as possible for all i, j
2. for $i = 1$ to N
3.   for $j = 1$ to f_i
4.   select integer s_{ij} as close to L/f_i as possible
5. for $i = 1$ to $N - 1$
6.   assign copies of i to the broadcast cycle positions with the objective
   of matching the equal-spacing criterion
7. assign copies of object N to the remaining free positions in the cycle

Figure 3.4. Wong's algorithm.

EXAMPLE 2 *We shall use the objects from Example 1 to illustrate Wong's algorithm. We can determine that all objects in a partition are broadcast in the same frequency with objects in partition A being broadcast approximately 6 times more often than those in partition C and objects in partition B being broadcast approximately 3 times more often than those in partition C. By fixing the length of the broadcast cycle to 60 objects, we have the resultant broadcast cycle as shown in Figure 3.5. We note that all the objects are equally spaced in this example.* □

14	20	27	11	69	35	45	29	22	31	14	50	60	65	69	25	33	29	75	85
29	22	31	14	17	38	47	69	25	33	29	52	62	14	20	27	67	69	80	88
69	25	33	29	19	41	14	20	27	48	69	55	63	29	22	31	14	72	82	91

Figure 3.5. Non-flat program generated by Wong's algorithm.

Vaidya and Hameed's online algorithm [119]. Vaidya and Hameed [119] showed that the equal-sized spacing of multiple copies of objects is also optimal for variable sized objects. Moreover, it was shown that the access time is minimized when the spacing between copies of objects is proportional to $\sqrt{l_i}$ and inversely proportional to $\sqrt{q_i}$, i.e., $s_i^2 \frac{q_i}{l_i} = constant$ where copies are equally spaced with spacing s_i. Based on this observation, Vaidya and Hameed proposed an *online* algorithm that picks the next object to broadcast as follows. Let T denote the current time and $R(j)$ denote the time at which a copy of object j was most recently transmitted. Then, at time T, object i is broadcast if $(T - R(i))^2 q_i/l_i \geq (T - R(j))^2 q_j/l_j$ for $1 \leq j \leq N$.[1] The algorithm is shown in Figure 3.6. We note that the expression $(T - R(i))^2 q_i/l_i$ is similar to $s_i^2 \frac{q_i}{l_i} = constant$. This algorithm produces a cyclic program if ties are resolved deterministically. However, the size of the cycle may be large, and hard to predict accurately.

Input: access probabilities; $q_1, \ldots, q_N$, object sizes, $l_1, \ldots, l_N$;
time to broadcast the next object, T
Output: the next object to be broadcast at T

for $i = 1$ to N
  $F(i) = (T - R(i))^2 q_i/l_i$
choose object i such that $F(i) = \max_{j=1}^{N} F(j)$
broadcast object i at time T
$R(i) = T$

Figure 3.6. Vaidya and Hameed's online algorithm.

EXAMPLE 3 *Figure 3.7 illustrates part of the broadcast program obtained from the Vaidya and Hameed's online algorithm for our running example. We note that though all objects appear once within a cycle of length 54 (i.e., smaller than Wong's scheme), the objects do not appear equally spaced.* □

14	27	29	91	33	82	80	14	69	20	63	33	31	25	27	47	20	41	33	31	11
17	69	22	88	85	31	75	27	67	29	62	60	55	50	14	69	45	38	14	69	29
11	25	20	14	69	29	72	25	22	65	14	69	52	48	25	22	29	35	19	17	27

Figure 3.7. Non-flat program generated by Vaidya and Hameed's online algorithm.

Several other methods have been proposed in the literature. We shall briefly summarize them here:

- In [119], an offline algorithm that generates cyclic non-flat program was also proposed. This algorithm essentially applies the online algorithm for a certain fixed broadcast length to obtain a broadcast sequence. Objects are then broadcast based on this sequence repeatedly. Hence, a cyclic program is obtained. However, the minimum length must be set such that each object must appear at least once in the cycle.

- In [52], it was observed that the generation of a broadcast program is similar to the problem of *packet fair queuing* [20, 98]. The result led to an adaptation of a packet fair queuing algorithm in [20] for broadcast scheduling. The resultant algorithm has log-time complexity ($O(log(N))$) and was shown to achieve near-optimal performance.

- Su and Tassiulas [103] have also addressed the issue of broadcast scheduling. They observed that two factors can influence the choice of an object i being picked for dissemination. The first is the elapsed time, w_i, since the last transmission of the object. The second is the rate of requests for the object, λ_i. A family of algorithms called *priority index policies* was proposed to pick the object with the maximum $\lambda_i^\gamma w_i$ value as the next object to be broadcast, where γ is a parameter that determines the relative importance of the two factors. In their performance study, it was shown that the best overall performance is obtained when $\gamma = 0.5$.

Each of the above algorithms is shown to be near-optimal when compared to the lower bound or optimal program in their respective works. The online algorithm was also shown to perform better than the probabilistic approach by a factor of 2 [119]. Regrettably, there is no comparative study amongst the various near-optimal schemes.

Generating non-flat coalesced broadcast programs

As mentioned, the main drawback of non-flat broadcast programs is the high computational cost (typically of order $O(N)$). Several efforts have opted to generate sub-optimal non-flat coalesced broadcast programs that incur much lower computational cost (time complexity of the order of $O(k)$ where k is the number of partitions).

Broadcast Disks [2]. Acharya, et. al. [2] were among the first to study coalesced broadcast program. In their *broadcast disks* approach, the data file is split into n *partitions*, where each partition comprises objects with similar access frequencies. Partitions with larger access frequencies will be broadcast more often than those with smaller access frequencies. Let there be k partitions and partition i be broadcast λ_i times. Moreover, let $\lambda_i > \lambda_j$ for $i < j$ and $\lambda_k = 1$. Let $\lambda = least\ common\ multiple\ of\ \lambda_i$. Each partition is further split into c_i *chunks* ($c_i = \lambda_i/\lambda$). The data file is then broadcast by interleaving chunks from the various partitions using the algorithm in Figure 3.8.

Input: partitions $P_1, \ldots, P_k$; λ; number of chunks for the partitions, $c_1, \ldots, c_k$
Output: cyclic non-flat coalesced program

```
for i = 0 to λ − 1
    for j = 1 to k
        broadcast chunk (i mod c_i) of partition P_j
```

Figure 3.8. Broadcast disks' interleaving strategy.

The broadcast can also be viewed as a sequence of equal sized *segments*, where each segment contains $\sum_{i=1}^{n} \|partition_i\|/\lambda_1$ objects ($\|partitions_i\|$ denotes the number of objects in partition i).

EXAMPLE 4 *From the data objects given in Example 1, we decide to broadcast objects in partitions A, B and C 4 times, 2 times and 1 time in a broadcast cycle respectively. Therefore, objects in partition B are split into two chunks, while those in partition C are split into four chunks. The resulting broadcast is shown in Figure 3.9, with a broadcast cycle of length 48 only.* □

A	B1				B2										
14	20	11	35		27	45	50			60	65			75	85
29	22	17	38	A	31	47	52	A	B1	62	67	A	B2	80	88
69	25	19	41		33	48	55			63	72			82	91

segment 1 | segment 2 | segment 3 | segment 4

Figure 3.9. Non-flat coalesced program generated by broadcast disks.

Vaidya and Hameed's bucketing algorithm [119]. Vaidya and Hameed [119] have also proposed an online algorithm with *bucketing* to generate non-flat coalesced programs. The algorithm splits the database into k partitions,

$P_1, P_2, \ldots, P_k$ such that P_i contains n_i objects and $\sum_{i=1}^{k} n_i = N$. Each partition is also associated with a queue. At any time, only an object at the front of a queue (partition) will be broadcast. Let d_j denote the average length of objects in partition P_j, i.e., $d_j = (\sum_{i \in P_j} l_i)/n_j$. Let a_j denote the average probability that an object in P_j is requested, i.e., $a_j = (\sum_{i \in P_j} q_i)/n_j$. Let T be the current time when an object has to be picked for broadcast, and $R(i)$ denote the time when object i was most recently broadcast. Let I_j denote the object at the front of the queue of partition P_j. It was shown that the minimum access time is achieved when the spacing for copies of an object i is proportional to $\sqrt{d}_j/\sqrt{a}_j$, i.e., $s_i^2 d_j/a_j = constant$. Based on this result, the proposed algorithm, shown in Figure 3.10, picks the object with maximum $(T - R(I_j))^2 d_i/a_i$ value as the next object to be broadcast. We note that every object in a partition has the same broadcast frequency. There is no guarantee that the program is cyclic.

In [119], the bucketing algorithm is compared against the online algorithm. As expected, it performs worse. In fact, the bucketing algorithm performs poorly when there is only one partition, but its performance improves with increasingly larger number of partitions.

Input: access probabilities, $q_1, \ldots, q_N$; object sizes, $l_1, \ldots, l_N$;
time to broadcast the next object, T
Output: the object to be broadcast at time T

1. for $i = 1$ to k
2. $\quad F(i) = (T - R(I_i))^2 d_i/a_i$
3. choose partition P_i such that $F(i) = \max_{j=1}^{k} F(j)$
4. broadcast object I_i at the front of the queue of P_i at time T
5. dequeue I_i and enqueue it at the rear of the queue of P_i
6. $R(I_i) = T$

Figure 3.10. Vaidya and Hameed's bucketing algorithm.

EXAMPLE 5 *Figure 3.11 illustrates part of the broadcast program obtained from the Vaidya and Hameed's bucketing algorithm for the same example in Example 1. Again, we note that objects do not appear equally spaced.* □

Discussion. Comparing the two methods for generating coalesced programs — broadcast disks vs bucketing — we observe the following. First, the broadcast disks ensures that all copies of an object are equally spaced. This is realized at the expense of "holes" (or idle periods when no object is broadcast).

11 17 69	20 19 14	35 22 29	38 41 45	69 25 27	14 47 31	33 29 48	50 52 69	20 55 14	22 60 62	29 63 65	25 69 27	67 14 31	33 72 29	75 80 82	20 69 85	22 88 91	29 25 11	27 17 69

Figure 3.11. Non-flat coalesced program generated by Vaidya and Hameed's bucketing algorithm.

The bucketing approach avoids the holes altogether but objects are no longer guaranteed to be spaced equally. Second, broadcast disks is designed for fixed-sized objects. It is not clear as to how it can be adapted to deal efficiently with variable-sized objects. For example, a large object may have to be split across multiple chunks which will complicate the generation of the program. The bucketing mechanism has no constraint on the size of objects, and has been shown to work well with variable-sized objects. Third, the effectiveness of the broadcast disks approach depends on the rate at which each disk "spins", and the issue of determining the optimal frequencies for the different disks is treated as an orthogonal issue to the construction of the broadcast program. The bucketing algorithm, however, integrates the two issues together, i.e., the algorithm attempts to optimize the performance of the program by taking the optimal frequencies into consideration. Finally, while the broadcast disks generates cyclic programs, the bucketing approach provides no such guarantee. The regularity of cyclic programs facilitates the incorporation of meta-data to facilitate *selective tuning*. Finally, it is interesting to note that in [2], it was advocated to have as few partitions as possible (to minimize complexity in the generation of program for broadcast disks). However, the results in [119] clearly shows that more partitions are preferred as more partitions can lead to better performance.

3.1.3 Generating Broadcast Program for Range Queries

The methods that we have seen so far optimize access to a single object. Interestingly, none of the non-flat broadcast programs can outperform the flat broadcast program (provided keys are sorted) for range queries. Since objects in non-flat broadcast programs are not sorted, the entirety of the "cycle" has to be examined. One straightforward approach, called NBP, works as follows [114, 115]: (a) generate a coalesced broadcast program using the broadcast disks technique, (b) sort the objects within every segment, (c) provide an additional range indicator at the beginning of each segment. Sorting objects within a segment can reduce energy consumption since clients can doze off if they find that their desired ranges in the segment have been broadcast, and only operate in active mode when the next segment is broadcast. Unfortunately, this method remains inefficient when dealing with range queries. This is because the ranges of any two arbitrary segments may overlap, and hence multiple segments may have to be searched to perform a range query (even when the results can be found in a single segment). Consider the running example and the program generated by broadcast disks in Figure 3.9. From Figure 3.9, we note that, even if the objects within each segment are sorted, all the four segments overlapped in their ranges (the ranges for the four segments are 11-69, 14-69, 14-72 and 14-91 respectively). As such, searching for objects in the range 30-40 may require scanning all the four segments!

To facilitate range queries, Tan and Yu [114, 115] proposed a novel algorithm, called NrBP, that was shown to handle range queries effectively without sacrificing much on the advantages of non-flat broadcast for single object retrievals. The basic idea of the algorithm is (a) to organize the data into partitions (as is done in the broadcast disks approach), (b) determine the number of times each partition should be broadcast, (c) sort the original data file and range partition the sorted file into disjoint and non-overlapping key ranges, (d) depending on the number of times each object is to be broadcast, the remaining copies are then replicated across the segments. The algorithm (for step (d)) is shown in Figure 3.12. As shown, the sorted data file, S, is split into λ_1 segments (line 1). The range indicators are assigned to the ranges of the respective segments (line 2). Recall that partition P_i has to be broadcast λ_i times $(1 \leq i \leq \lambda_1)$. At line 1, the segments only contain one copy of every partition P_i. Therefore, the remaining number of copies for partition P_i to be broadcast becomes $(\lambda_i - 1)$. In order to ensure that single object retrievals can also be effectively handled, copies of the same object should be allocated as equidistant from one another as possible. Since it is not always possible to do so, a four-step heuristic approach was adopted. In the first step, the allocation is based on the sorted data file, S, which only contains a single copy of every partition P_i (line 1). In the second step, as P_i is broadcast λ_i times, the interval (line 5) between two copies of the object in partition P_i is estimated as $|S|/\lambda_i$. In the third step, for each object p in a partition P_i to be replicated (line 6), its initial position within S can be determined using the function determinePos() which is essentially a binary search algorithm. Finally, in the last step, the remaining $(\lambda_i - 1)$ copies of the object p in the partition P_i are replicated to the respective segments (lines 8-11). The segments in which the $(\lambda_i - 1)$ copies of object p have to be kept are determined by the heuristic function determineSeg() as follows: the position of an object, k, is used as a hint to determine the next copy of the object in S. Based on the object key, the segment is determined using the partitioning vector, v. Once all the records are allocated, each segment is sorted (line 15). Finally, the program is then the concatenation of the various segments together with their respective ranges (line 16).

EXAMPLE 6 *Figure 3.13 illustrates the algorithm for the running example in Example 1. In Figure 3.13(a), the objects are range-partitioned into four segments with the ranges 11-27, 29-47, 48-65 and 67-91 respectively. A range*

```
Input: sorted data file, S; partitions P_1, P_2, ..., P_n
          where P_i is broadcast λ_i times (λ_i ≥ λ_j, for i < j)
Output: A non-flat broadcast program, BP

split sorted data, S, into λ_1 segments, s_1, ..., s_{λ_1}, of equal size;
let range_i be the range indicator for segment s_i, 1 ≤ i ≤ λ_1;
let v[λ_1 − 1] be the resulting range partitioning vector;
for i = 1 to n − 1 do
    interval = |S|/λ_i;
    for all p ∈ P_i do
        k = determinePos(p, S);
        for j = 1 to λ_i − 1 do
            k = k + interval;
            m = determineSeg(k, v, S);
            s_m = s_m ∪ {p};
        endfor
    endfor
endfor
sort each segment;
BP = [(range_1, s_1), (range_2, s_2), ..., (range_{λ_1}, s_{λ_1})]
```

Figure 3.12. Algorithm NrBP.

query that requests for objects in the range 30-40 only needs to access objects in the second segment. Moreover, processing at the client can terminate as soon as object 41 has been scanned. Figure 3.13(b) shows the remaining objects that are to be replicated across the various segments — since partition A is to be replicated 4 times, only 3 copies of its objects are left; similarly, only 1 copy of objects in B needs to be replicated, while objects in partition C have already been allocated. Figure 3.13(c) shows the allocation of the remaining objects to the various segments. For example, since object 29 belongs to A and has already appear in segment 2, the other copies are allocated to segments 1, 3 and 4. Similarly, object 20 has already appeared in segment 1, and so it is being allocated to segment 3. Finally, Figure 3.13(d) gives the resultant broadcast. As can be seen from the figure, not all segments are of the same length. Segment 1 has 10 objects while segment 3 has 15 objects. Some load-balancing mechanisms can be introduced to ensure that all segments are of the same length, but the average access time is not expected to be significantly different as the length of the broadcast cycle will not be affected. □

In the simulation study conducted in [107], three broadcast programs are studied: the non-flat coalesced broadcast programs generated by NrBP and

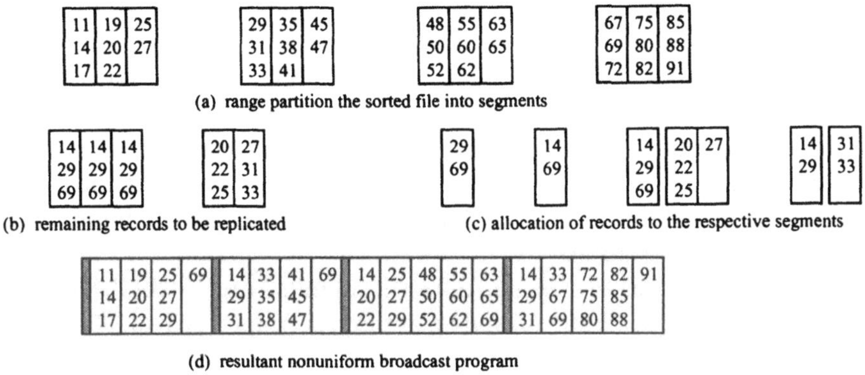

Figure 3.13. Non-flat coalesced broadcast program generated by algorithm NrBP.

NBP, and the flat broadcast program. The results showed that both the non-flat broadcast programs perform equally well in terms of single object retrievals, and outperform the flat broadcast program by a wide margin. However, for range queries, the flat broadcast program performs the best, followed by the non-flat program by NrBP, and finally, the program by NBP. The study also showed that, unless the percentage of range queries is high, the program generated by NrBP performs best overall.

3.2 SCHEDULING STRATEGIES FOR PULL-BASED BROADCAST SYSTEM

In a pull-based broadcast system, there is an uplink channel available for the clients to submit object requests. Thus, the server knows the exact number of pending requests for each object, and can determine the next object to broadcast based on that information. While the scheduling algorithms designed for the push-based broadcast system can be employed for pull-based broadcast system, they are not expected to perform well for several reasons. First, in the push-based approach, the access probabilities of all objects are fixed; here, the requests on different objects change over time (in fact, whenever an object is broadcast, the access probabilities differ). Second, in push-based systems, the schedules can be constructed offline; but offline schedules are not expected to perform well for pull-based systems. Third, the scheduling decision is more critical for pull-based systems. For example, picking an irrelevant object to disseminate can result in a sharp increase in the number of uplink requests from many clients. As another example, when multiple clients submit requests for the same (popular) objects, transmitting multiple copies of the objects will lead to wastage and poor utilization of the wireless bandwidth. Fourth, as client population increases, the channel and the server may not be able to cope with the demand. Thus, the scalability problem remains an issue as the client population increases.

To overcome these problems, the server should "intelligently" transmit only one copy of the data objects to satisfy multiple requests that arrive within a short duration of time. This is possible with wireless medium since a single broadcast of an object satisfies all of the outstanding requests for the object. Thus, a good scheduling algorithm has the potential to greatly improve the effectiveness of the broadcast.

Essentially, when a client needs an object, it submits a request for the object to the server. Client requests are queued up at the server upon arrival. The server repeatedly chooses an object from among these requests, broadcasts it over the wireless medium, and removes the associated request(s) from the queue. Clients monitor the broadcast and receive the objects that they are waiting for. It is the *scheduling algorithm* employed by the server that determines the next object to be broadcast (and hence the order in which the requests are serviced). In this section, we shall look at scheduling algorithms that have been proposed in the literature. After examining the main challenge in developing scheduling algorithms, we review existing scheduling techniques on fixed-sized objects and variable-sized objects in Sections 3.2.2 and 3.2.3 respectively.

For the remaining of this section, we assume a fixed-length unit of data transfer between the server and the clients. We shall refer to these units as *pages*. Objects are organized into pages for broadcasting.

3.2.1 Performance Metrics

A scheduling algorithm must scale well. At the same time, there are three main criteria for evaluating scheduling algorithms for large-scale data dissemination:

- *Responsiveness.* An effective scheduling algorithm should provide quick response to user requests. There are two criteria along this dimension: the *average access time* measures the amount of time on average from the moment the data arrives at the server to the time that the answer is broadcast; and the *worst case access time* determines the maximum amount of time that any client request will have to wait before the answers are received. Being able to balance the two criteria is one main challenge in designing scheduling algorithms.

  For variable-sized objects, since different objects have different service time, the access time measure is thus considered to be biased towards small sized objects. As such, the notion of *stretch*, which is defined to be the ratio of the access time of a request to its service time, has been proposed as a fairer performance metrics for variable-sized objects [6]. The rationale for this choice is that the stretch of a request translates more directly to user-perceived performance — intuitively, clients requesting a large object should be expecting to have a longer access time than those requesting smaller objects.

- *Scheduling Overhead.* The overhead incurred by the scheduling algorithm is an important consideration. This is especially so since an algorithm with high overhead is not likely to scale. There are two types of scheduling overhead. The first concerns the processing of a new request when it arrives. For example, upon arrival, certain data structures may have to be updated, the request may have to be inserted into a queue, and so on. The second deals with selecting the next object to broadcast. If the decision incurs excessive overhead, the server may not be able to support a large number of requests. Typically, the two types of overhead are related: spending more effort when a request first arrives can reduce the cost of scheduling decisions.

- *Robustness.* The success of a scheduling algorithm depends on how well it adapts to changes in the workload as well as access patterns. Being robust to such changes is important since changes may lead to poor performance if some underlying assumptions no longer hold.

3.2.2 Scheduling Strategies for Fixed-Sized Objects

Scheduling Theory is a mature algorithmic area that has been extensively researched since the late 1960's [50, 69]. However, it is interesting to note that most of work are directed at unicast rather than broadcast scenario. We shall present some of the basic scheduling schemes for fixed-sized objects here [40, 121]. We note that for most of these schemes, the entirety of an object is broadcast without preemption. In other words, we can consider each object as having the size of a page. Some of these schemes are:

- *Earliest First Request (EFR).*[2] Under the EFR strategy, data objects are broadcast in the order of the arrival time of the first request. The scheme has the advantage that it is "fair" in that requests arriving early will be served first, and for those that "cut queue" (because their requested objects are the same as earlier requests) there is no penalty on the part of those that wait.

- *Most Requests First (MRF).* In MRF, the data object with the maximum number of pending requests is picked for broadcasting. Thus, the scheme favors popular objects and waiting time for these objects can be reduced at the expense of less popular objects. A variant of MRF is the Most Requests First Lowest (MRFL) scheme which breaks ties in favor of the page with the lowest access probability. We note that MRF-based schemes have to work hand-in-hand with some other heuristics to deal with requests on unpopular objects; otherwise, such requests may have very long waiting time or may never be served! One possible heuristic is to give priority to an object whose waiting time exceeds a certain tolerable value.

- *Longest Wait First (LWF).* The total waiting time for an object is the sum of the time that all pending requests for an object have been waiting. LWF selects the page that has the largest total waiting time. This scheme also favors popular objects since more requests will increase the total waiting time. However, requests on unpopular objects will not be "starved"; instead, they will be served eventually when it waits long enough.

In [7], a simulation study[3] was conducted to compare EFR, MRF, MRFL and LWF. The results showed that EFR performs best for unpopular objects but not for popular objects. On the other hand, MRF is superior for popular objects but performs poorly for unpopular objects. It turns out that the average access time of both schemes are poor. LWF, being able to treat both popular and unpopular objects more fairly, was shown to perform the best on average access time. In fact, it was observed that under very high load, all the schemes' average access time sort of flatten out (i.e., stays constant) with LWF outperforming EFR and MRF by up to 3 times. It was also noted that MRFL can outperform EFR for small databases, but is worse than EFR for large databases. This is because, in large databases, the opportunity for tie breaking diminishes and MRFL degenerates to MRF.

While LWF provides good average case performance, it suffers from high overhead: whenever an object has to be broadcast, LWF recalculates the total accumulated wait time for every object with pending requests in order to decide which object to broadcast. Thus, it is likely to become a bottleneck for a system with a large database, making it impractical for such systems.

A promising alternative scheduling algorithm with low overhead was proposed in [7]. The algorithm, called RxW (Request times Wait), combines the benefits of MRF and EFR in a way that ensures scalability by having low overhead. RxW schedules the object with the maximal $R \times W$ value where R is the number of outstanding requests for that object and W is the amount of time that the oldest of those requests has been waiting for that page. Thus, RxW schedules an object for broadcast either because it has many outstanding requests or it has at least one request that has waited for a long time.

To avoid computing the RxW values for all objects, the algorithm maintains two sorted lists to prune the search space— one ordered by R values and the other ordered by W values (W can be obtained from the difference between the current clock and the arrival time of the oldest request.) The two lists are threaded through the service queue, which has a single entry for any requested object of the database. The procedure for finding the maximal $R \times W$ value works as follows. From the object at the top of the R list, the corresponding W value can be determined. This W value can then be used to compute the limit for possible W values since the object with the maximum RxW value cannot have a W value below this limit. Next, the entry for the object at the top of the W list is used to limit the R value. This process alternates between the two queues until the limit is reached on one of them, in which case, the maximum RxW valued object is obtained.

A parameterized variant of RxW was also proposed in [7], denoted $RxW(\theta)$, to further cut down the overhead. $RxW(\theta)$ employs a first-fit heuristic: instead of picking the maximum RxW valued object, $RxW(\theta)$ selects the *first* object whose RxW value is greater than or equal to $\theta \times threshold$ where $threshold$ is the running average of the RxW value of the last object that was broadcast and the $threshold$ at that time. θ is essentially a tuning knob that facilitates the trading off of scheduling overhead and average wait time performance. The smaller the value of θ, the fewer objects will be scanned. By setting θ to 0, the algorithm degenerates to comparing only the two top entries from the R and the W lists in determining the object to be broadcast. If there is no object with RxW value greater or equal to $\theta \times threshold$, then the algorithm acts like the regular RxW algorithm and returns the object with the maximum RxW value.

The experimental study reported in [7] showed that RxW performs almost as well as LWF, making it a truly low overhead alternative for LWF. The study also showed that a θ value of 0.9 can also cut down on the overhead without much sacrifice on the average case performance. Even when θ is 0, the algorithm remains attractive. Moreover, it was shown that $RxW(\theta)$ has lower worst case performance for smaller θ value, while LWF and RxW performs poorly for worst case performance.

In [103], a parametrized algorithm called *priority index policies* (PIP-γ) was proposed. PIP-γ schedules the page with the maximum $\lambda_i^{-\gamma} X_i$ value where λ_i is the rate at which request for object i is generated, X_i is a measure of the number of request backlog for object i and γ is a weight used to determine the relative importance of the λ_i and X_i. While it was shown that PIP-0.5 performs equally well as LWF at low overhead, its assumed knowledge of the probability of access for each object restricted its applicability to stable environments.

3.2.3 Scheduling Strategies for Variable-Sized Objects

The scheduling strategies for fixed-sized objects can also be applied for variable-sized objects. However, variable sizes of objects open up opportunities for new mechanisms to be designed that are not meaningful for fixed-sized objects. For example, for variable-sized objects, the service time (the time to broadcast an object) can be used as a scheduling criterion, whereas all fixed-sized objects have the same service time. One of the earliest scheduling strategies for variable-sized objects is the *Shortest Service Time First (SSTF)* scheme. In SSTF, the data object with the shortest service time is chosen for broadcast. It is biased towards small objects, and requires some additional heuristics to deal with requests on very large-sized objects.

More recently, Acharya and Muthukrishnan argued that *preemptive* scheduling schemes are natural and beneficial for scheduling variable-sized objects [6]. In preemptive schemes, the broadcast of an object may be interrupted (preempted) to service other requests before resuming the remainder of the original broadcast (object). Preempting the broadcast of a large object to broadcast another object with a large number of outstanding requests can improve performance since more clients are satisfied earlier. This makes preemption applicable and useful for disseminating fixed-sized objects too.[4] However, besides additional complexity in the algorithm, preemption incurs the overhead of buffering more objects at the server. In [6], several preemptive scheduling schemes were proposed:

- *Preemptive Longest Wait First (PLWF).* PLWF is essentially the preemptive counterpart of LWF. In other words, after the broadcast of a page, the LWF criterion is employed to determine another (may be the same) object from which a page will be picked for broadcast. Like the LWF scheme, PLWF has high scheduling overhead.

- *Preemptive Remaining Service Time (PRST).* PRST is the preemptive version of SSTF scheme: once a page of an object is broadcast, the SSTF criterion is applied to pick the next page from a subsequent object for broadcast. Interestingly, PRST (also known as the *Shortest Remaining Processing Time*) is shown to optimize the average access time in point-to-point connection [13].

- *Longest Total Stretch First (LTSF).* Under LTSF, the page from the object with the largest total current stretch is picked for broadcast. The total stretch is given by the sum of the current stretches of all pending requests for the object.

- *MAX algorithm.* The MAX algorithm is a low overhead approximation of the computationally expensive BASE algorithm [18]. MAX first computes the maximum stretch for any object. This is given by the maximum value of the individual stretch of all satisfied requests. This maximum stretch value is then used to set a deadline for incoming requests. When requests arrive, or are completed, the maximum stretch value may be updated. However, once the deadline for a request is set, it is not changed even if the stretch value is updated before the request is serviced. The object with the *earliest deadline first* will be the candidate object for broadcast.

In [6], a simulation study was conducted to evaluate PLWF, SRTF, LTSF and MAX, The results showed that none of the schemes perform best in all cases: PLWF is superior in terms of worst case access time, LTSF performs well for average access time, SRTF outperforms the others in average stretch performance, and MAX performs generally well in terms of both average access time and stretch (both worst and average case).

Notes

1 Initially, $R(j) = -1$ for $1 \leq j \leq N$, and $T = 0$.

2 In [6, 7], this scheme was termed First-Come-First-Serve. We felt that this term does not reflect the scheme accurately since requests that arrive at a later time may also be served as long as the object that is broadcast (in response to the request at the front of the queue) is their targeted object.

3 The study did not consider the overhead associated with running the scheduling algorithm at the server.

4 This is applicable for large fixed-size objects whose size is larger than a page.

Chapter 4

INDEXING ON AIR

Portable devices that operate in wireless environments are typically powered by batteries. As a result, power conservation is a key issue [44, 72, 79]. Unfortunately, the lifetime of a battery is very short and is not expected to increase much [10, 96]. Moreover, when a client is listening to the channel, the CPU must be in active mode to examine the data packets. This operation is very expensive from an energy point of view, because often only few data packets are of interest to a particular client. It is therefore important for a mobile client to be able to operate under two different modes: *doze mode* where it is still connected to the network but it is not active; and *active mode* where it performs its usual activities. Clearly, an important topic is to provide mechanisms that allow clients to switch from doze mode to active mode alternatively when listening to the broadcast data. By selectively tuning to the desired portion of the broadcast, energy consumption can be minimized and the effective battery life can be prolonged.

A large part of this chapter is devoted to provide a comprehensive review of techniques that "index" the broadcast data to facilitate selective tuning for push-based model of information dissemination. However, we shall devote the last section to look at selective tuning techniques that have been designed for pull-based model.

4.1 DATA ORGANIZATION FOR SELECTIVE TUNING

We assume that the server organizes the data to be broadcast into a broadcast program. The database is broadcast to the clients according to the program on the downlink channel (also called *broadcast channel*). Clients receive the broadcast data and filter them.[1] Filtering is performed by a simple pattern matching operation against the key value. Thus, clients can remain in the doze mode most of the time and tune in periodically to the broadcast channel to

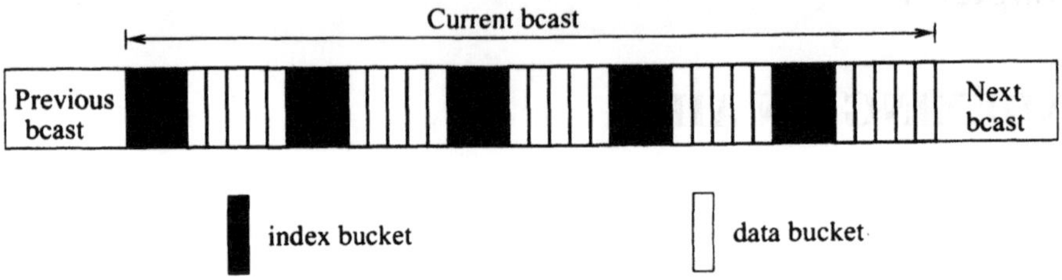

Figure 4.1. Basic broadcast organization with indexes.

download the required data. To allow clients to *selectively tune* to the broadcast, the clients need to be equipped with a *directory* that indicates the point of *time* in the broadcast channel when particular objects will be broadcast. There are two solutions to this:

1. *The mobile clients cache a copy of the directory.* This solution, however, has limited applicability. Since mobile clients may move from cell to cell, it means that the clients will have to cache all the directories of all the cells that they may possibly move to. The storage requirement on the portable clients may become too high. In addition, the clients must frequently refresh their copies of the directories to ensure consistency. This may also generate excessive traffic between clients and the server.

2. *The directory is broadcast in the form of an index on the broadcast channel.* In this case, the client must first access the directory once it moves into a cell. This approach incurs the overhead of additional channel bandwidth requirement for broadcasting the directory, but it is scalable and avoids the problems that are associated with the caching approach. Thus, this is the most widely used and preferred solution.

Figure 4.1 shows a general organization for broadcast data (including the directory). Each broadcast *version* of the file, together with all the interleaved index information, is called *bcast*. A *bcast* consists of a certain number of buckets, each representing the smaller unit that can be read by a client (thus, a bucket is equivalent to the notion of block for disk organizations). Pointers to specific buckets are specified as an *offset* from the bucket containing the pointer to the bucket to which the pointer points to. The time to get the data pointed by an offset s is given by $(s-1) \times T$, where T is the time to broadcast a bucket.

Figure 4.2 shows the general protocol for retrieving broadcast data:

1. The mobile client tunes to the channel and looks for the offset pointing to the next index bucket. During this operation, it must be in active mode. Most works assume that each bucket contains the offset to the next index bucket. Thus, this step requires only one bucket access. Let n be the determined offset.

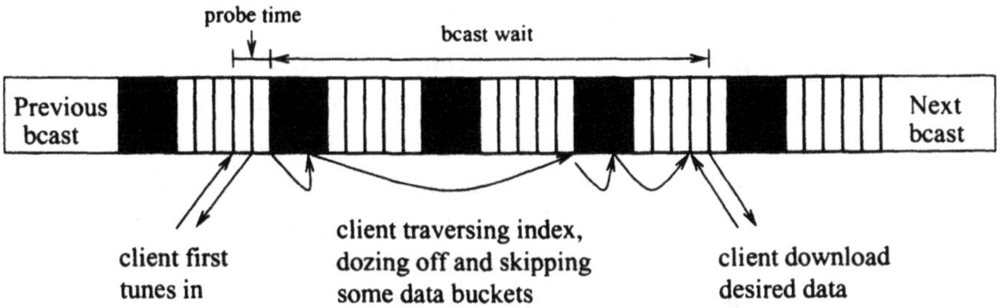

Figure 4.2. Basic protocol for retrieving broadcast data.

2. The client switches to the doze mode until time $(n - 1) \times T$. At that time, it tunes into the channel (thus, it is again in active mode) and, following a chain of pointers, determines the offset m, corresponding to the first bucket containing the data of interest (with respect to the considered key value). We note that during this process, the client may switch between the doze mode and active mode multiple times.

3. The client switches to doze mode until time $(m - 1) \times T$. At that time, it tunes into the channel (thus, it is again in active mode) and retrieves the data of interest.

In general, any existing indexing structures can be adapted as a directory. However, since the wireless medium is a sequential medium, the issue is to be able to multiplex the index and the data effectively. Two metrics have been used to measure this effectiveness:

- *Access time*: It is the average duration from the instant in which a client wants to access objects with a specific key value to the instant when all required objects are downloaded by the client. The access time is based on the following two parameters (see Figure 4.2)

  – *Probe time*: The duration from the instant in which a client wants to access objects with a specific key value to the instant when the nearest index information related to the relevant data is obtained by the client.

  – *Bcast wait*: The duration from the point the index information related to the relevant data is encountered to the point when all required objects are downloaded.

  Note that if one parameter is reduced, the other increases.

- *Tuning time*: It is the time spent by a client listening to the channel. Thus, it measures the time during which the client is in active mode and therefore determines the power consumed by the client to retrieve the relevant data.

We note that the two metrics conflict. The use of a directory reduces the tuning time, but it lengthens the access time. On one extreme, if we do not employ an index, the access time is minimal while the tuning time is the worst (i.e., tuning time is as large as the access time). On the other extreme, by broadcasting a single directory at the beginning of the data, we obtain the best tuning time, but higher access cost. It is therefore important to determine good bucket interleaving strategy in order to obtain a good trade-off between access time (thus reducing the time the client has to wait for relevant data) and tuning time (thus reducing battery consumption).

4.2 FLAT BROADCAST PROGRAMS WITH INDEXES

In this section, we shall look at indexing schemes for flat broadcast programs. As mentioned, existing indexing schemes in conventional databases can be adapted for broadcast programs. In particular, the techniques that are based on the three main categories of indexing schemes, namely, tree-based, hash-based and signature-based, have been explored.

4.2.1 (1,m) Index

The (1,m) indexing scheme [60] is an index allocation method where a complete index is broadcast m times during a *bcast* (see Figure 4.3). All buckets have an offset to the beginning of the next index segment. The first bucket of each index segment has a tuple containing two fields. The first field contains the key value of the object that was broadcast last and the second field is an offset pointing to the beginning of the next *bcast*. This tuple guides clients who missed the required object in the current *bcast* so that they can tune to the next *bcast*.

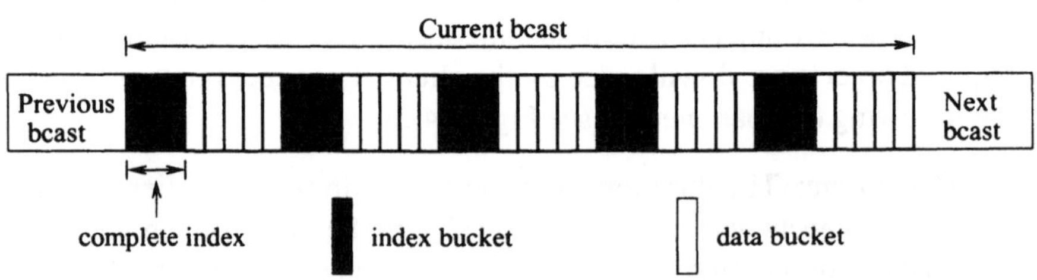

Figure 4.3. Bcast organization in the (1,m) indexing method.

Assuming multi-level indexes, the client's access protocol for retrieving objects with key value K is as follows:[2]

1. Tune into the current bucket on the broadcast channel. Get the offset to the next index segment.

2. Goto the doze mode and tune in at the broadcast of the index segment.

3. Examine the tuple in the first bucket of the index segment. If the target object has been missed, obtain the offset to the beginning of the next *bcast* and goto 2; otherwise, goto 4.

4. Traverse the index and determine the offset to the target data bucket. This may be accomplished by successive probes, by following the pointers in the multi-level index. The client may doze off during the process (between two probes).

5. Tune in when the desired bucket is broadcast, and download the bucket (and subsequent ones as long as they contain the key K).

The main problem with the (1,m) index organization is that the entirety of the index has to be replicated m times. This increases the length of the broadcast cycle and hence the average access time. The optimal m value that gives minimal average access time is shown to be $\sqrt{\frac{data\ file\ size}{index\ size}}$. The scheme has, however, good tuning time.

We note that there is actually no need to replicate the complete index between successive data blocks. Rather, it is sufficient to make available only the portion of index related to the data buckets which follow it. This is the approach adopted in all the subsequent indexing schemes that we will be looking at.

4.2.2 Tree-based Index

In the tree-based indexing scheme (also referred to as the *distributed indexing* scheme in [60]), a data file is associated with a B^+-tree index structure [33] (see Figure 4.4). Since the broadcast medium is a sequential medium, the data file and index must be *flattened* so that the data and index are broadcast following a *preorder traversal*[3] of the tree. Furthermore, the index comprises two portions: the first k levels of the index will be *partially* replicated in the broadcast, and the remaining levels will not to be replicated. The index nodes at the $(k+1)$th level are called the *nonreplicated roots*.

Essentially, each index subtree whose root is a nonreplicated root will appear once in the whole *bcast* just *in front of* the set of data segments it indexes. In other words, each nonreplicated root and its descendant nodes in the index will appear *only once* in a given version of a *bcast*. On the other hand, the nodes at the replicated levels are replicated at the beginning of the *first* broadcast of each of its children nodes. (*first* because its children nodes may be replicated also.) Thus, the number of replications of each node at the replicated levels is determined by the number of children it has. Referring to Figure 4.4 as an example, we note that node a_1 will be replicated three times — at the beginning of b_1, b_2 and b_3. Similarly, the root node will be replicated only at the first a_1, first a_2, and so on. The resultant organization of the broadcast for the example is shown in Figure 4.5.

To facilitate selective tuning, each node contains meta-data that help in the traversal of the trees. All non-replicated (index and data) buckets contain pointers that will direct the search to the next copy of its replicated ancestors. For example, data buckets 1, 2 and 3 will provide an offset to the second a_1. On the other hand, all replicated index buckets contain two tuples that can direct the search to continue in the appropriate segments. The first tuple is a pair (x, ptr_{begin}) that indicates that key values less than x have been missed and so the search must continue from the beginning of the next *bcast* (which is ptr_{begin} buckets away). The second pair (y, ptr) indicates that key values greater than or equal to y can be found ptr offset away. We note that x and y are respectively the minimum and maximum values in the data segment rooted by the index bucket. As an example, the first a_1 will contain the pairs (1, offset to first I) and (81, offset to second I). Clearly, if the desired object has key value between x and y, the search can continue as in conventional search operation.

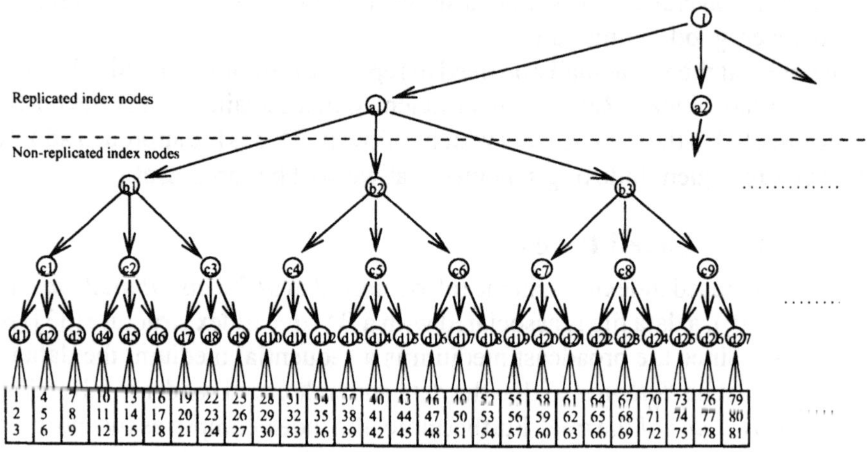

Figure 4.4. A partial data file and its index tree.

The access protocol of this scheme for a client searching for an object with key K is as follows:

1. Tune to the current bucket of the *bcast*. Get the offset to the next index bucket, and doze off.

2. Tune to the beginning of the designated bucket and examine the meta-data.

   - If the desired object has been missed, doze off till the beginning of the next *bcast*. Goto 2.

   - If the desired object is not within the data segment covered by the index bucket, doze off to the next higher level index bucket. Goto 3.

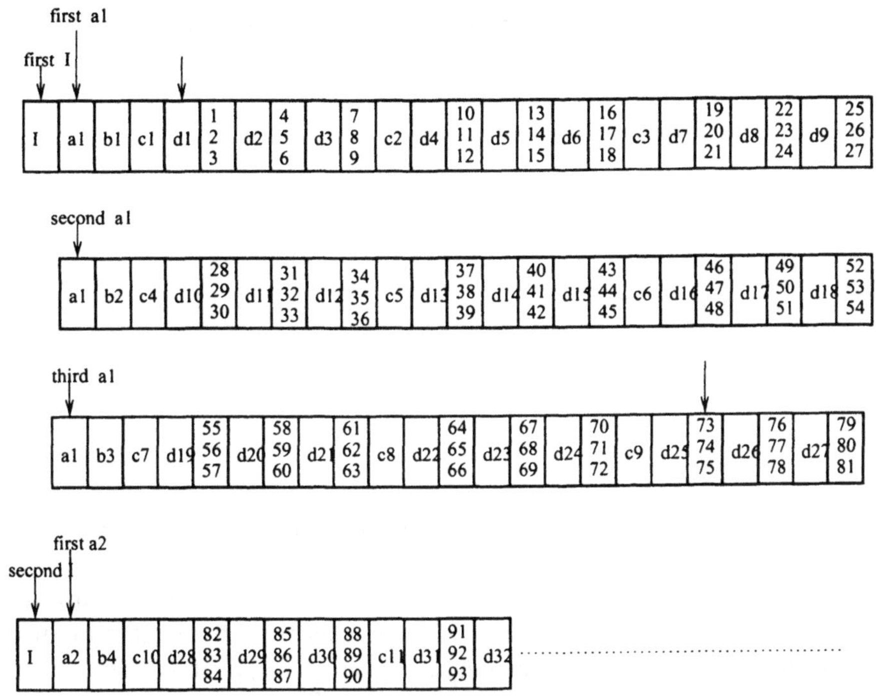

Figure 4.5. The data broadcast for the distributed indexing scheme with partial path replication.

- If the desired object is within the data segment covered by the index bucket, goto 3.

3. Probe the designated index bucket and follow a sequence of pointers to determine when the data bucket containing the target object will be broadcast. The client may doze off in between two probes.

4. Tune in again when the bucket containing objects with key K is broadcast, and download the bucket (and all subsequent buckets as long as they contain objects with key K).

Comparing with the (1,m) index, the tree-based scheme has lower access time since it has a shorter broadcast cycle (as not the entirety of the index is replicated). Its tuning time is also comparable to that of the (1,m) scheme.

4.2.3 Hash-based Index

In a hash-based scheme, data are hashed into a set of partitions. Because of the nonuniform distribution of data, partitions may differ in size. Partitions are further organized into fixed size buckets, i.e., each partition consists of one or more buckets with the last bucket being possibly partially full. The data are then broadcast in partitions as a sequence of buckets. Figure 4.6 shows an example of 4 partitions. In the figure, each box represents a bucket. An unfilled box represents the beginning of a partition. As shown, partitions 1, 2, 3 and 4 have 4, 6, 3 and 3 buckets respectively.

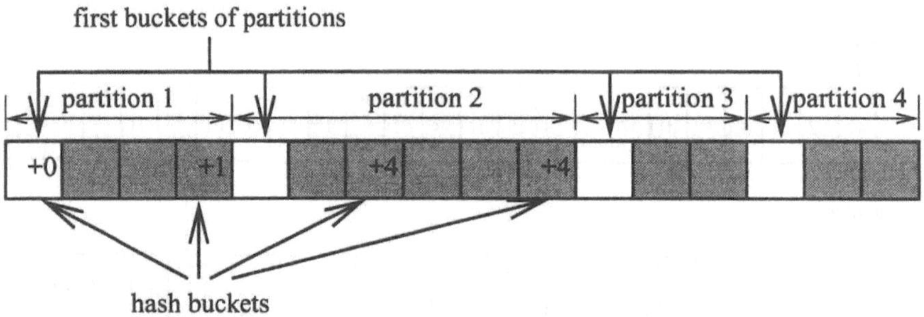

Figure 4.6. A broadcast with 4 partitions.

To facilitate selective tuning, every bucket contains some control information: a *bucket id* which is the offset of the bucket from the beginning of the version, two *hash functions* (denoted $h_{partition}$ and h_{hash}) [4] and a *pointer* which is the offset to the beginning of the next version. The motivation for two hash functions will become clear shortly. From this information, a client can doze off and tune to the beginning of the next version if it has missed the desired partition. Furthermore, for a file with *N* partitions, *N* (predetermined) buckets contain additional control information: a *shift pointer*. We shall refer to these *N* buckets as the *hash buckets*. The reason for introducing the hash buckets is because it is not possible to determine the offset of the desired partition since partitions may be of unequal length and the broadcast channel is a sequential medium. The hash buckets thus provide an indirect addressing mechanism to the partitions. The two hash functions are as follow: for a given key value K, $h_{partition}(K)$ determines the partition that object K belongs to, and $h_{hash}(K)$ determines the hash bucket that contains the shift pointer to the corresponding partition. We note that $h_{hash}(K) = f(h_{partition}(K))$. In other words, to search for an object with key K, we determine the partition it belongs to (using $h_{partition}$). From this, we can determine its hash bucket (using h_{hash}). The shift pointer, being stored in the hash bucket, can then be retrieved to provide the offset (from the hash bucket) to the actual address of the desired partition.

We shall focus on the *Hashing B* protocol in [61], which is shown in Figure 4.6. Basically, in this technique, the hash buckets are spaced out evenly (except for the last hash bucket). The *gap* between hash buckets is given by the size of the smallest partition in terms of number of buckets. For example, in Figure 4.6, each hash bucket is equally-spaced 3 buckets away since the smallest partition contains 3 buckets. In other words, h_{hash} is of the form

$1 + (h_{partition} - 1) \cdot gap$. As shown, the shift pointers for the various buckets indicate the offsets to the start of the corresponding partitions. For example, the shift pointer for partition 1 is +0 which means that partition 1 begins at the hash bucket. On the other hand, the shift pointer in the hash bucket for partition 4 is +4 which indicates that partition 4 is 4 buckets away. This *equi-spacing* hash method has been shown to outperform the simple way of assigning the first N consecutive buckets in a broadcast as hash buckets (which is equivalent to having a gap of 1).

Let the search key be K. The client's access protocol is as follows:

1. Tune to the current bucket, and compute $h_{partition}(K)$ and $h_{hash}(K)$. Let the hash bucket be h_i. Let the bucket id of this current bucket be bid.

2. Determine N_{doze}, the number of buckets to doze off before hash bucket h_i will be broadcast. Note that there are two cases to consider in determining the number of buckets to doze off:

   - *The target hash bucket will be broadcast in this cycle soon.* If $bid \leq h_i$, then $N_{doze} = h_i - bid$. For example, if $bid = 4$ and h_4 is the target hash bucket, then $N_{doze} = 9 - 4 = 5$. Goto 3.

   - *The target hash bucket in this cycle has been missed.* If $bid > h_i$ then N_{doze} is given by the offset to the beginning of the next version of *bcast*. For example, if $bid = 10$ and h_2 is the target hash bucket, then $N_{doze} = 6$. Goto 1.

3. Obtain the *shift* value. Switch to doze mode until *shift* number of buckets.

4. Examine the objects in the partition bucket by bucket for the desired object.

The hash-based scheme requires minimal meta-data, and so is expected to perform well in terms of access time. However, it has to examine the entirety of a partition to find the objects. This may incur high tuning time especially for large partitions.

4.2.4 Signature-based Index

Signature techniques have been successfully employed to facilitate speedy retrieval in conventional databases [29], text databases [42], image databases [31] and multimedia databases [85]. A signature is basically an abstraction of the information stored in an object or a file. It is typically formed by hashing each value in the object into a bit string and then superimposing all bit strings generated from the object together to obtain the object signature, say S_i. During filtering, a query signature, S_q, is constructed in the same manner and then compared to the signatures of all objects. An object signature matches the query signature if for every bit that is set in the query signature, the corresponding bit is set in the object signature; otherwise, the object does not match the query. Clearly, it is possible for false drops to occur, i.e., an object signature and a query signature match but the object does not satisfy the query. However, the signature-based approach offers several advantages that makes it a promising index structure for wireless environments [77]. First, signature techniques can be applied to all kinds of information media. Second, signatures are easy to generate and search with small buffer requirement. Third, signatures are compact encodings of the data they represent. In addition, the length of the signature can be controlled by varying the number of values hashed into a signature (with the corresponding effect on the number of false drops). Finally, a signature file (containing all signatures of the objects) is a sequential file structure. This makes it easy to "linearize" the signature file for broadcasting on air and scanning by a mobile client, as compared to a tree-based approach which lose its speed advantage since random access cannot be done on a broadcast channel.

There are several signature-based techniques. On the one hand, each object[5] can be used to construct a signature (called *simple signature*). On the other hand, we can group several objects to produce an *integrated signature*. In the former case, we have the finest granularity and as many simple signatures as there are objects. The latter case has coarser granularity, and the degree of coarseness can be controlled depending on the number of objects that are grouped. In [77], the multi-level signature-based approach [74] was adapted to facilitate selective tuning in wireless context. The scheme comprises k levels of signatures for some $k > 1$. The lowest level, level k, comprises simple signatures. Each upper level comprises integrated signatures as follows: the integrated signatures at level $k - 1$ are obtained from a group of objects at level k, the integrated signatures at level $k - 2$ are obtained by combining multiple groups from level $k - 1$, and so on. The higher the level is, the coarser the granularity of the grouping is. The broadcast is organized such that each integrated signature is broadcast before the corresponding group of objects, and each simple signature is broadcast before the corresponding object. Thus, each integrated signature indexes all the objects between itself and the next integrated signature at the same or higher level. To minimize complexity, the signatures at all levels are typically of the same size. Moreover, to reduce the number of false drops, the hashing functions used in generating the signatures at different levels should be different. We note that with each signature, there is also additional information maintained. These include the offset to the next signature at the same level or at a higher level, and the level of the signature. Figure 4.7 illustrates an example of a 3-level signature-based approach. Here, the integrated signatures at level 2 are obtained from grouping 2 objects, and those at level 1 are obtained from grouping 4 objects.

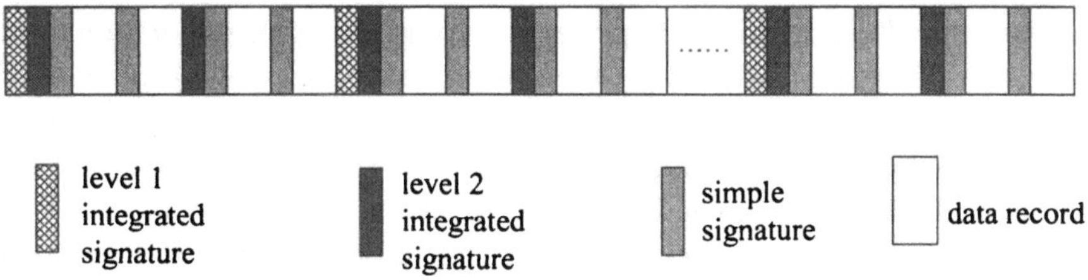

Figure 4.7. Signature scheme.

To answer a query, a query signature is generated for each level of the signatures. Let the query signatures at level i be Q_i. The protocol to access objects of interest is as follows:[6]

1. Tune to the current bucket and listen for a signature.
2. Upon receipt of the signature, compare it with the corresponding query signature at that level.
3. If it is an integrated signature
   (a) If the query and integrated signatures do not match, determine when the next integrated signature at the same or higher level will be broadcast; doze off for that amount of time and goto Step 2.
   (b) If there is a match, we need to look out for signatures at the next (lower) level. If it is not the lowest level, goto Step 2; otherwise, goto Step 4;
4. If it is a simple signature
   (a) If the query and simple signatures do not match, determine when the next simple signature or integrated signature will be broadcast; doze off for that amount of time and goto Step 2.
   (b) If there is a match, the corresponding object is received, and further check to see if it satisfies the query. Read the next signature. Goto Step 2.

Because signatures are compact, the tuning time is expected to be small. However, since there is no way to know whether an object exists or not unless the entirety of the signature file has been "scanned" (not necessarily examined), the access time is at least one broadcast cycle.

4.2.5 Flexible Indexing Scheme

The flexible indexing scheme, proposed in [61], splits a sorted list of objects into several equal-sized segments, and provides indexes to navigate through the segments. This is achieved as follows. At the beginning of each segment, there is a *control index* which comprises two components: a *global index* and a *local index*. The global index is used to determine the segment which an object may be found, while the local index provides the offset to the portion within the segment where the object may be found.

Suppose the file is organized into p segments. Then the global index at a segment, say s, has $\lceil log_2 i \rceil$ (*key*, *ptr*) pairs, where i is the number of segments in front of and including segment s, *key* is an object key, and *ptr* is an offset. For the first entry, *key* is the key value of the first data item in segment s and *ptr* is the offset to the beginning of the next version. By examining this pair, the client will know if it has missed the data, and if so, it will wait till the broadcast of the next version of the file. For the jth entry ($j > 1$), *key* is the key value of the first data item in the $(\lfloor \frac{log_2 i}{2^{j-1}} \rfloor + 1)$th segment following segment s and *ptr* is the offset to the first data bucket of that segment.

The local index consists of m (*key*, *ptr*) pairs that essentially partition each segment further into $m + 1$ sections. For the first entry, the *key* value is the the key of the first data item of section $m + 1$ and the *ptr* is the offset to that section. For the jth pair, *key* is the key value of the first data item of section $(m + 1 - j)$ and *ptr* is the offset to the first data bucket of that section.

From the above description, it is clear that the number of segments, and the number of sections per segment can affect the performance of the scheme. Now, increasing the number of segments or sections will increase the length of the broadcast cycle and reduce the tuning time, and vise versa. Thus, the scheme is flexible in the sense that it can be tuned to fit an application's needs.

Figure 4.8 illustrates the flexible indexing scheme. Here, the file is divided into 8 segments of 4 objects each (except for segment 8). Consider segment 2. There are 3 entries in the global index. The first entry, (12, 27), implies that any object that is less than 12 can be found after the 27th object, i.e., in the beginning of the next version of the file. The second entry, (54, 13), means that an object larger than or equal to 54 can be found after the 13th object, i.e., in segment 5. Similarly, the third entry (42,5) indicates that an object between 42 and 54 may be found 5 entries away, i.e., two segments away. Implicitly, objects that may be found in the segment follows the local index. The local index has 3 entries. Consider the local index of segment 2. The entry (29, 4) means that an object with value 28 and above may be found after 4 objects away. The entry (22, 3) refers to an object with value above 22 but less than 29 can possibly be found 3 objects away. Likewise, the entry (20, 2) points to where objects between 20 and 22 can be found. Implicitly, any object less than value of 22 can be found at the end of the local index.

Given a search key, K, the access protocol is as follows:

1. Tune into the channel for a bucket, obtain the offset to the next index segment. Doze off until the index segment is broadcast.

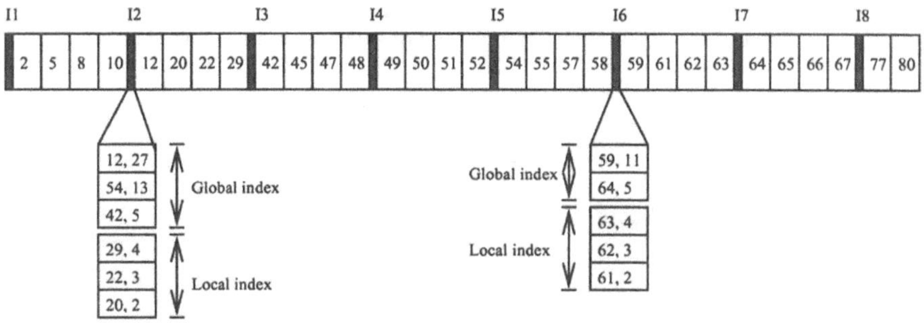

Figure 4.8. Flexible index scheme.

2. Examine the global index entries. If the target object belongs to another segment, get the offset, doze off for the appropriate amount of time and goto 2.

3. Examine the local index entries. Obtain the offset to the section where the target data is stored. Switch to doze mode for amount of time determined by offset.

4. Examine objects in the data bucket for the desired object, and download the object.

4.2.6 Discussion

Table 4.1 summarizes the qualitative performance of the various selective tuning schemes that we have seen so far. As shown, we expect the $(1,m)$ scheme, the tree-based technique and the signature file method to perform poorly in terms of access time. This follows from the fact that both the $(1,m)$ scheme and the tree-based method have the longest cycle length because of their meta-data's size overhead is the most significant. The signature-based scheme performs poorly since it requires a full cycle to find an object. While the hash-based scheme has a shorter cycle length than the flexible indexing technique, its access time performance is moderate because clients tend to miss the hash bucket in their initial probe (i.e., the probe time is high). The flexible scheme provides the shortest access time.

In terms of energy consumption, the $(1,m)$ scheme, the tree-based scheme and the flexible indexing scheme all have low tuning time. The signature based scheme is moderate because of the false drops that it has to handle. Finally, the hash-based scheme is expected to be the worst as it requires the CPU to stay active to examine objects of a partition once the partition is determined (it is unable to selectively tune within a partition).

Table 4.1. Qualitative comparison of selective tuning schemes.

Scheme	**Access time**	**Tuning time**
(1,m)	poor	good
tree-based	poor	good
hash-based	moderate	poor
signature-based	poor	moderate
flexible	good	good

4.3 SELECTIVE TUNING FOR SKEWED DATA ACCESS

When some objects are more frequently accessed than others, the access frequencies can be exploited to design index methods that further minimize the average number of index probes (as compared to the techniques that we have seen so far). One approach, which we shall defer its discussion to the next section, is to design index methods for non-flat broadcast programs. An alternative approach, which we shall discuss here, is based on flat broadcast programs. The basic idea is to construct an imbalanced tree index where frequently accessed data are placed "closer" to the root node, while unpopular data are placed "further" from the root node. Since energy consumed is proportional to the number of index nodes examined, this arrangement effectively minimizes the average amount of energy consumed by reducing the number of probes for frequently accessed data at the expense of unpopular data. We shall look at one such algorithms proposed by Chen, Yu and Wu [30], algorithm *CF* (*constant fanout*).

Algorithm CF is a generalized version of Huffman code algorithm [71]. As shown in Figure 4.9, it adopts a bottom-up strategy to create an imbalanced tree of fixed fanout d by repeatedly merging d subtrees (except for the last merge which may have fewer than d subtrees) with the smallest total access probabilities. For a given fanout d, the average number of index probes obtained from the tree constructed by algorithm CF is optimal. However, it is not clear how to obtain the optimal fanout that can give the optimal performance.

EXAMPLE 7 *Figure 4.10 illustrates algorithm CF. Here, we have nine objects, o1 to o9, represented by square boxes. The internal nodes are represented by circles. In the figure, the label besides each node is the access probability for the node. Figure 4.10(a) illustrates the tree index obtained with a fanout of 3, and Figure 4.10(b) shows the corresponding flat broadcast program.*

Input: access probabilities of each object; desired fanout d
Output: flat broadcast program with index

```
for i = 1 to N
    create a subtree with only object i as its node
    each subtree has an associated access probability of p_i
n = N
while n > 1
    merge the f = min(n, d) nodes with the smallest access probabilities
        into a single node (i.e., a new node with d children nodes)
    let the access probabilities of the new node be the sum of the
        probabilities of all its children
    n = n - f + 1
flatten the tree to obtain the broadcast program
```

Figure 4.9. Algorithm CF.

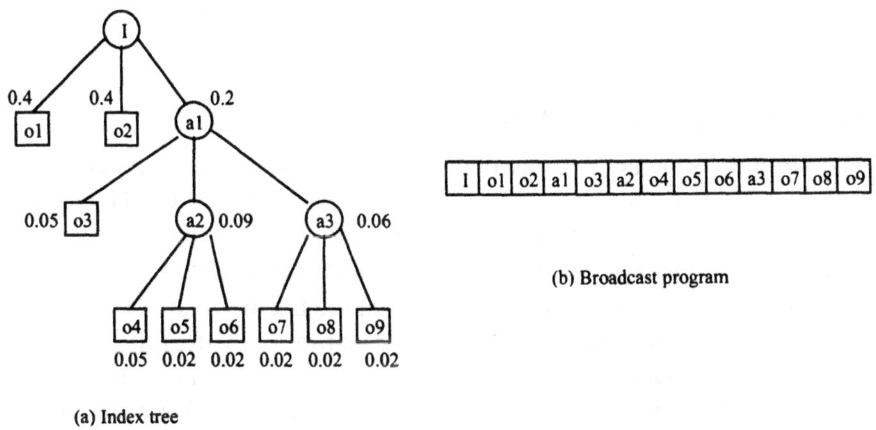

Figure 4.10. Imbalanced tree index for skewed data access.

4.4 NON-FLAT BROADCAST PROGRAMS WITH INDEXES

We have seen how non-flat broadcast programs can help to minimize the average access time and how selective tuning has been employed for flat broadcast programs. Unfortunately, selective tuning methods for flat broadcast are not directly applicable for non-flat broadcast. For example, consider the most straightforward method of indexing a broadcast file, where the index is broadcast immediately before the data file. Building such an index on non-flat broadcast is meaningless since the benefits of the duplicated objects cannot be exploited. As another example, hashing techniques cannot be employed at all since it is unlikely that there exists a hashing function that maps a single object into multiple partitions where copies of the object are replicated. Moreover, existing indexes do not account for multiple copies of the same objects. In addition, existing indexes assume objects are ordered by some key values. For flat broadcast programs, such an ordering can be exploited to reduce the tuning time. For example, given a key value, it can be easily determined where the object is stored. However, in a non-flat broadcast, such an ordering cannot be exploited in a straightforward manner. This is because in a non-flat broadcast, objects that are broadcast frequently will have overlapping ranges with those that are not frequently broadcast. In other words, objects in a non-flat broadcast are only, at best, partially sorted. Therefore, to exploit selective tuning for non-flat broadcast, new indexing mechanisms must be devised. We shall look at several techniques proposed by Tan and Yu [111, 126]. These techniques operate in two phases. In the first phase, a non-flat coalesced program is constructed using the broadcast disks method. Recall that the program can be seen as comprising a sequence of segments. In the second phase, a known selective tuning method is applied on the program by treating each segment as a flat program.

4.4.1 Segment-Level Index

The simplest scheme is to associate an index with each segment of the broadcast. This is similar to the (1,m) index scheme. In the (1,m) index scheme, the entire broadcast forms a segment. In non-flat broadcast, the entire broadcast is split into several segments, with overlapping key ranges.

The basic idea is to broadcast a *full* index at the beginning of each segment, i.e., the index contains entries for the entire data file. Only the full index at the *beginning of the broadcast* contains offsets to *all* the objects *directly*. The other indexes contain both *direct* and *indirect* offsets. Direct offsets are used for objects that can be accessed within the same broadcast. Indirect offsets point to the index at the beginning of the next *bcast*, and are used for objects that can only be accessed in the next *bcast*. Figure 4.11 illustrates the full index non-flat broadcast program. There are 33 objects divided into three partitions: partition A has 6 objects (chunks A1, A2), partition B has 9 objects (chunks B1, B2, B3), and partition C has 18 objects (chunks C1-C6). Here, the broadcast program is generated under broadcast disks with partitions A, B and C being broadcast 3 times, 2 times and once in a broadcast cycle respectively. In the figure, the full index contains entries for all objects in partitions A, B and C. The index, F1, contains all direct offsets – offsets to objects of A1, A2, B1 to B3, and C1 to C6. Tuning into F1 will allow the client to doze off until the desired object appears. On the other hand, F2 contains direct offsets for partitions A (in segment 2) and B (in segments 2 and 3), and chunks C3 to C6 (in segments

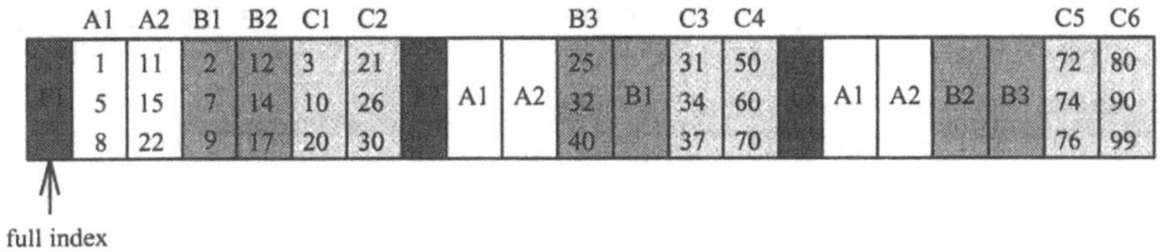

Figure 4.11. Full index scheme.

2 and 3), but indirect offsets to C1 and C2 (in segment 1). Here, a client that tunes into F2 to access an object in C1 must also tune into F1 again before the object can be retrieved.

Under this strategy, the access protocol is as follows:

1) Tune into the downlink channel to determine the address of the next index segment.

2) Go into the doze mode and tune in at the broadcast of the index segment. From the index segment, determine whether the desired object exists. If the object is not found, then the object does not exist in the data file. Otherwise,

   - for a direct offset, determine when the data bucket containing the desired object is broadcast. This is accomplished by successive probes and by following the pointers in the index; go into doze mode.
   - for an indirect offset, determine the offset (which is for the first index at the beginning of the next broadcast), go to step 2.

3) Tune in when the bucket containing the desired object is broadcast and download the object.

The segment-level index scheme has a problem: the cost to find the index may be very large. Referring to Figure 4.11, if the client tunes in at A1 in the first segment, and it wants to retrieve an object in A2, it must first doze off and wait for F2 before finding the object. The time to wait for F2 may be large depending on the size of the segment. To cut down on this "additional" index wait time, the index may be replicated several times and distributed within each segment. While one can find an index faster now, the price is a larger broadcast cycle as a result of more copies of indexes being broadcast.

4.4.2 Distributed Indexing

To reduce the long access times of segment-level schemes, each segment index can be "split" into sub-indexes that are distributed within its corresponding segment. This is the approach adopted in the *distributed indexing scheme*. The distributed indexing scheme adapts and extends the flexible indexing scheme for non-flat broadcast. Recall that the flexible indexing scheme splits a sorted list of objects into several equal-sized sections. At the beginning of each section, there is a control index which comprises two components: a global index and a local index. The global index is used to determine the section which an object may be found, while the local index provides the offset to the portion within the section where the object may be found. In flat broadcast, it suffices to look into a section/portion to determine whether an object exists or not. However, in non-flat broadcast, because ranges of keys overlap between segments, when an object is not found in a section/portion, it does not mean that it does not exist. Therefore, when an object is not found, other segments have to be searched as well. The distributed indexing scheme works as follows:

- Organize the non-flat broadcast into segments
- Sort the objects in each segment
- Build a flexible index for the objects in each segment. This is done by splitting a segment into a number of *sections*, and each section into a number of *portions*. For section i of a k-section segment, its global index contains $k - i + 1$ entries that provide offsets to all sections that come after it. Its local index will have m pointers that split the data in the section into $m + 1$ portions and provide the offsets to the $m + 1$ portions.

Figure 4.12 illustrates the distributed indexing scheme using the same set of objects in Figure 4.11. Here, each segment is divided into 3 sections of 6 objects each. Consider segment 2. For its first index, there are 3 entries in the global index. The first entry, (1, 18), implies that any object that is less than 1 can be found after the 18th object, i.e., in the next segment. The second entry, (34, 12), means that an object larger than or equal to 34 can be found after the 12th object, i.e., in the last section of the segment. Similarly, the third entry (11,6) indicates that an object between 11 and 34 may be found 6 entries away, i.e., in the second section of the segment. Implicitly, objects that may be found in the first section of the segment follows the local index. The local index has 2 entries. Consider the local index for the first index of segment 2. The entry (8, 4) means that an object with value 8 and above may be found after 4 objects away. The entry (3, 2) refers to an object with value above 3 but less than 8 can possibly be found 2 objects away. Implicitly, any object less than value of 3 can be found at the end of the local index.

Under this scheme, a timeout is necessary to identify an object that does not exist. The access protocol for the distributed indexing scheme (assuming object is in the broadcast) is as follows:

1) Tune into the downlink channel to determine the address of the next index segment.

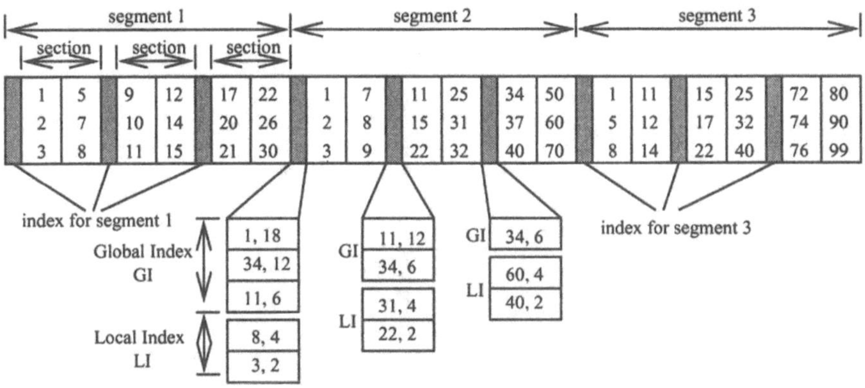

Figure 4.12. Distributed index scheme for non-flat broadcast program.

2) Go into the doze mode and tune in at the broadcast of the index segment. From the index segment, examine the global index. If the object may be found in the other segment, get the offset to the index of the next segment; go to step 2. Otherwise

   - If the object may be found in the current section, examine the local index and obtain the offset to the respective portion; go to doze mode.
   - If the object may be found in another section, get the offset to the index of the respective section, go to step 2.

3) Tune in when the bucket containing the desired portion is broadcast

4) Examine the objects in the portion. If the object is not found, get the offset to the index of the next segment; goto step 2.

4.4.3 Discussion

We present here a set of experimental result to show the relative performance of the segment-level index scheme (FI) and the distributed index scheme (DI). The non-flat broadcast program without index (NI) is also used as a reference. The experiment evaluated the effect of different frequency of broadcast for each partition (i.e., the rate at which each broadcast disk "spins") by varying the control parameter, δ. Figure 4.13 shows the result when δ varies from 1 to 7. The results show several interesting observations. First, there is a certain δ that leads to optimal performance. This is true for all schemes studied, be it a no indexing scheme or an indexing scheme.

Second, NI performs the best in terms of access time but has the largest tuning time. This is expected since NI has the shortest broadcast cycle and the client has to listen and examine all objects until the desired object is found.

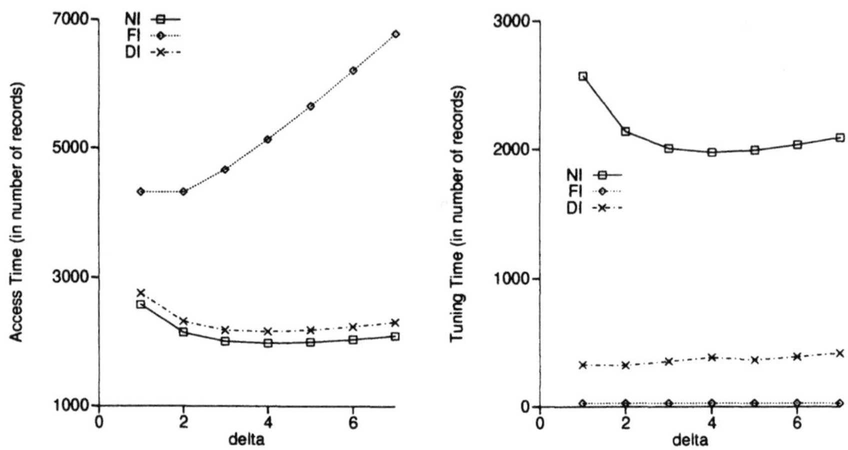

(a) Access time. (b) Tuning time.

Figure 4.13. Effect of δ.

Third, we note that the access time for FI is the worst, while its tuning time is the best. The high access time of FI is contributed by two factors: the index size and the time to wait for the index segment. Intuitively, on average, a client will have to wait for half the segment length before the index segment is probed. For small δ value, the segment size becomes large, and so the access time will also be large. For large δ, the number of copies of the index also increases. This again leads to longer broadcast cycle and hence higher access time. Its tuning time is low since only one, and at most two indexes need to be examined.

Finally, the DI scheme provides a good compromise. Its access time is at most 120% that of NI, but its tuning time is almost 10 times less than NI. This implies that DI can reduce the tuning time without significant increase in access time.

4.5 SELECTIVE TUNING MECHANISMS FOR PULL-BASED MODEL

As noted, selective tuning mechanisms are largely studied in the context of push-based environment. Such schemes, unfortunately, cannot be applied to pull-based context because demand-driven data cannot be predetermined and vary from query to query. To facilitate selective tuning for pull-based environments, mechanisms are needed to allow clients to repeatedly toggle between *doze* mode and *active* mode until the resultant objects are obtained. In the active mode, clients will listen to the broadcast for up to t_a time units for the objects. In the doze mode, clients will "rest" for up to t_d time units. There are two categories of mechanisms in which these can be achieved [112]:

- **Stateful Server.** Schemes belonging to this category have the following characteristics:
  - clients determine the schedule for doze-off/wake-up, i.e., the t_a and t_d time units;
  - once clients submit their requests, they doze-off/wake-up according to the schedule;
  - the server is made known of the clients' schedules; this can be achieved by submitting the schedule together with the query;
  - the server follows the schedules of clients in responding to their requests, i.e., it will only download objects of clients during the periods in which they are active, and avoid transmitting objects during the doze off periods. Note that the server transmits the IP-address of the client, together with the resultant objects.
- **Stateless Server.** Schemes belonging to this category have the following characteristics:
  - after submitting query requests, clients look out for *cues* from the server;
  - server transmits *cues* together with data objects; the *cues* essentially indicate the content of the data that follows the *cues* and the time slot in which the next set of *cues* will be broadcast;
  - based on the *cues*, a client can doze off if its requests cannot be met, and wake up only when the next set of *cues* is broadcast.

Clearly, under the stateful approach, the client has control over the schedule in which it dozes off/wakes up. But, the server has to be very sophisticated and complex. The server must maintain and keep track of the schedules of all clients that have submitted queries. In addition, there may be several exchanges (or negotiations between clients and server) before a final "optimal" schedule can be determined. On the other hand, the stateless approach is simple, though it requires clients to abide by the cues of the server.

4.5.1 Stateful Selective Tuning Scheme

In a stateful selective tuning scheme, there are several issues that need to be addressed:

- How does the client determine its schedule?
- How does the server schedule the queries?

- What measures must be taken to ensure that the clients will eventually receive the results?

Tan and Yu [112] proposed a simple strategy that employs simple heuristics to handle the above issues. The scheme works as follows:

- The ratio of the amount of time spent in active and doze mode is fixed at t_a/t_d; once the query is submitted (t_a and t_d are piggy-backed to the query), the client will be active for t_a time units, after which it will doze off for t_d time units, and the process repeats.

- The server examines query requests using a FIFO policy. When a query is examined, from its t_a and t_d values and its arrival time, t_{arr}, the server is able to *predict* whether the corresponding client is in active or doze mode. Let $t_{ad} = t_a + t_d$, and $t_c = t_{arr} - \epsilon_c$. Here, t_c is an estimation (by the server) of the time that the client submits its query, and ϵ_c is the delay between the submission of the query by the client to the receiving of it by the server. This corresponds to the transmission time of the query over the uplink channel. k is determined as the smallest integer that satisfies the expression:

  $$t_c + k \cdot t_{ad} < t_{now} < t_c + (k+1) \cdot t_{ad}$$

  where t_{now} denotes the current time. The client is expected to be in active mode at time t_x if t_x satisfies the expression:

  $$t_c + k \cdot t_{ad} - \epsilon_s < t_x < t_c + k \cdot t_{ad} + (t_a - \epsilon_s)$$

  and ϵ_s is used to account for the delay before the client receives the IP-address (otherwise, the client may miss the IP-address although it has been broadcast). If the client is not in active mode, then the next query will be examined, and the process repeats.

- The client is also assigned a timeout period, $t_{timeout}$, so that if the client still has not received its first result after $t_{timeout}$ time units, then it will resubmit its query. This is to avoid the client from missing the data (especially since the client and server may not be synchronized).

The access protocol of the client is as follows:

1) Submit a query (or list of referenced objects) and a schedule;

2) Tune into the downlink channel, and listen for t_a time units;

   a) Examine the objects broadcast;

   b) If the query is being served (i.e., IP-address is in the broadcast), then download all resultant objects and STOP;

3) Go into doze mode and wake up after t_d time units;

4) If timeout, goto 1; otherwise goto 2.

Under this scheme, once the IP-address is broadcast, all the data objects required by the client with the associated IP-address are also broadcast. The scheme, however, fails to exploit much data sharing. For example, a client may have requested a subset of the data that is being broadcast, but it is unable to access them (since the objects of a query are identified by the IP-address). This will also lead to a long initial access time.

We would note that the above scheme is a *static* scheme in that once t_a and t_d are determined, they cannot be changed. Clearly, an optimal $[t_a, t_d]$ pair depends on the load of the system (i.e., number of clients in the mobile environment). When the number of clients is small, the access time of a query is expected to be low. In this case, it is more suitable to set a small doze off time or a large active time. On the other hand, if the number of clients is large, the access time is expected to be high, and so the doze off time can be larger to save up on the tuning time.

4.5.2 Stateless Selective Tuning Schemes

For the stateless schemes, the basic protocol between clients and server is as follows:

- Besides broadcasting the data objects, the server also broadcasts *cues*. The *cues* essentially provide hints to the content of the data that follows them. We also include the time slot in which the next set of *cues* will be broadcast at the end of the *cues*.

- Clients submit query requests. Moreover, clients also listen to the *cues*. From the *cues*, clients can determine whether the data objects being broadcast in the segment are their resultant objects. If the data objects are for them, then they will listen to the segment; otherwise, they will doze off for the period of time that is determined by the length of the segment (derived from the time slot in which the next set of *cues* is to be broadcast). In this way, clients can toggle between the doze mode and active mode repeatedly until all the resultant objects are received.

In some sense, the scheme is similar to the indexing schemes used in push-based approaches. The *cues* can be considered as a form of indexes, and the dissemination of data as segments, where each segment comprises a set of *cues* and its associated data objects (see Figure 4.14). However, unlike the push-based approaches, depending on the number of clients in the system and the size of the query results, segments are of variable length. Moreover, the content of the segments is also different and unpredictable.

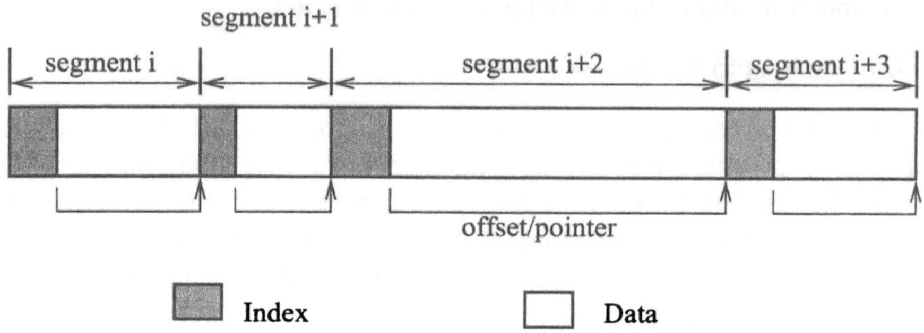

Figure 4.14. A broadcast to facilitate selective tuning.

Clearly, depending on the *cues* used, different selective tuning strategies can be derived. We shall look at two schemes proposed in [112].

Client Identity as Cues

The first strategy uses the client identity as the *cue*. Since the IP-address of a client is unique, it is used as the client identity. Each segment then comprises:

- A list of IP-addresses;
- An offset that indicates when the next segment will be broadcast;
- All the resultant data objects of clients whose IP-addresses are being broadcast.

This strategy includes all the resultant data for a query in a single segment. The specific access protocol for a client is as follows:

1) Submit a query (or list of referenced objects);

2) Tune into the downlink channel, and listen for the beginning of the next segment;

3) Read the $offset$ to determine the address of the next segment;

4) Read and examine the list of IP-addresses;

    a) If the client's IP-address is being broadcast, examine the data objects that are broadcast and download the referenced objects;

    b) If the client's IP-address is not being broadcast, go into doze mode and tune in at the broadcast of the next segment (based on the offset); goto 3.

Like the stateful-based scheme, this scheme cannot exploit much data sharing, since all the data objects are broadcast together with the IP-address in the same segment. The same object can appear multiple times in a segment. Hence, it is expected to have a long initial access time also.

Clearly, the strategy is a *multicasting* and *batching* mechanism. The batch size is a factor that can affect the effectiveness of the strategy. If the number of clients batched is small, then it may be repeatedly switching from one mode to the other. This may be bad when the number of resultant data objects per client is small, in which case, clients will be waking up almost immediately when they doze off. On the other hand, too large a number of clients may also be bad as clients in the segment may be listening to all the objects in the broadcast, resulting in a longer tuning time too.

Another factor that affects the performance of the strategy is the number of clients in the system. When the number of clients is small, it makes no sense to have a large batch of clients in the segment since the access time will be excessively large due to the additional waiting time for the required number of clients to arrive. However, for large number of clients, a large batch size may be useful as this will help to better utilize the channel bandwidth.

In [112], an adaptive approach was also proposed. The server determines the batching size based on the number of queries in the waiting queue and a predetermined *threshold* value (for the maximum number of queries to be batched). When the number of queries is less than the threshold, all the queries will be served in the current segment; otherwise, only the number corresponding to the threshold will be served. Figure 4.15 illustrates the scheme. Here, the threshold is set to four queries, i.e., at most four queries will be batched. At t_1, the waiting queue contains only two queries. As such, both queries will be served and their data objects transmitted in a single segment. At t_2 which corresponds to the time when segment 1 has completed and the time to transmit the next segment, the queue has increased to five queries. Since the threshold is set to four, we can only serve four queries. At t_5, there is no job in the queue. As a result, nothing needs to be transmitted. At t_6, a query arrives and is served immediately since it is the only job and the channel is available.

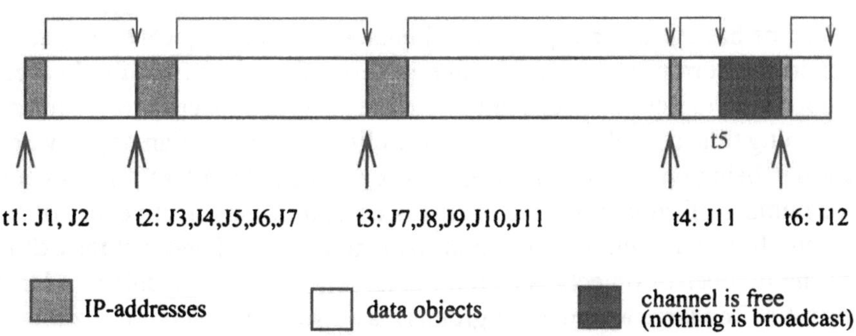

Figure 4.15. Client identity selective tuning scheme.

Data Groups as Cues

An alternative strategy is to partition a database into groups, and use the group-ids as the *cues*. The grouping function can be simply a modulo function, or can be different for different types of objects, or can be generated on-the-fly (such as using signatures). Whatever grouping function is employed, it must be agreed upon between the server and the mobile clients. In this case, the *cues* is a list of group-ids that indicate the groups that the data in the segment belong to. However, not all the objects in the group need to be broadcast. In other words, the group-ids provide only a hint on the objects that will be broadcast, but not the actual objects that will be broadcast. Clients will examine the group-ids; if their desired objects belong to the groups, it will listen to the broadcast, otherwise, it will doze off. Under this scheme, the access protocol for a client is as follows:

1) Submit a query (or list of referenced objects);

2) Tune into the downlink channel, and listen for the beginning of the next segment;

3) Read the $offset$ to determine the address of the next segment;

4) Read the list of group-ids, and examine them;

   a) If the objects referenced by the client's query belong to one of the groups being broadcast;

      i) Examine the data objects that are broadcast and download the referenced objects, if any;

      ii) If all the objects have been downloaded, then STOP; otherwise, doze off and tune in at the broadcast of the next segment (based on the offset), goto 3;

   b) If none of the groups broadcast are needed by the client, go into doze mode and tune in at the broadcast of the next segment; goto 3.

This scheme has the advantage that all clients whose data objects belong to the groups in the current segment can tune in (even though the required objects may not be in the current segment). However, it is expected to result in a larger tuning time as "false drops" (i.e., the client requests for an object whose group-id is being broadcast, but the object is not being broadcast) may occur.

The number of groups is a factor that is critical to the effectiveness of this approach. Too small a number of groups will increase the chances of false drops (since the number of objects per group is large), and hence result in a larger tuning time. Too large a number of groups may also result in poor utilization of the channel as more group-ids has to be transmitted. Like the IP-based scheme, the server determines the number of queries to batch adaptively (based on a *threshold* value). However, in this scheme, it is the group ids that form the *cues*, i.e., the group ids from all the resultant objects of the batched queries. Since different objects may belong to the same group, only one group-id for these objects need to be broadcast. Thus, the number of groups can vary from segment to segment.

Another factor that can affect the group-based scheme is the data access patterns. If the distribution of clients' accesses is uniform, then the false drop is expected to be small. On the contrary, if the distribution is skewed, then false drop may increase (especially when the group size is large).

4.5.3 Discussion

We evaluated the various pull-based selective schemes using a simulation model. Figure 4.16 shows a representative result showing the effect of query interarrival time on the strategies by varying the interarrival time from 0.1 to 1.0. The stateful-based scheme, the IP-based stateless algorithm and the group-based stateless algorithm are respectively denoted as SS, IP and GP. The basic pull-based strategy (denoted OD) is also used as a reference to study the benefits of the proposed schemes. Under scheme OD, clients listen to the broadcast until all data objects have been received (without dozing off at all).

From the results, we note that none of the strategies performs best in all cases. In terms of tuning time (Figure 4.16(a)), we see that all the proposed schemes (IP,GP and SS) are superior over scheme OD. This demonstrates the effectiveness of the schemes in minimizing power consumption. As expected, when the load is light, all the schemes require the same amount of tuning time (since the opportunity for dozing off is low). However, under high load, the tuning time of the proposed schemes is drastically reduced. For IP, its tuning time can be reduced to only 3% that of OD, while SS and GP can have tuning time of down to 20% and 30% that of OD respectively. GP is less effective because of high false drops.

For the average access time (Figure 4.16(b)), all the schemes perform equally well (or poorly) except for the stateful-based scheme (SS). Algorithm SS performs poorly (in access time) under lightly loaded context, but is superior when the system is heavily loaded. When the load is light, the access time is expected to be low. By allowing clients to doze-off for a long time (relative to the access time) inevitably increases the access time unnecessarily. However, as the system load increases, algorithm SS can lead to better average access time. While this appears unintuitive, three reasons account for this behavior. First, under other strategies (IP and GP), the cues are at the front of the segment, and every client's access time includes all the cues. But, under SS, objects that are broadcast earlier will have a smaller number of IP-addresses to examine. Secondly, the scheduling policy adopted minimizes the number of times a client needs to doze off since the next query to be addressed is examined in FCFS policy. Finally, under SS, some queries are processed earlier, while others are delayed. It turns out that the net effect is a gain.

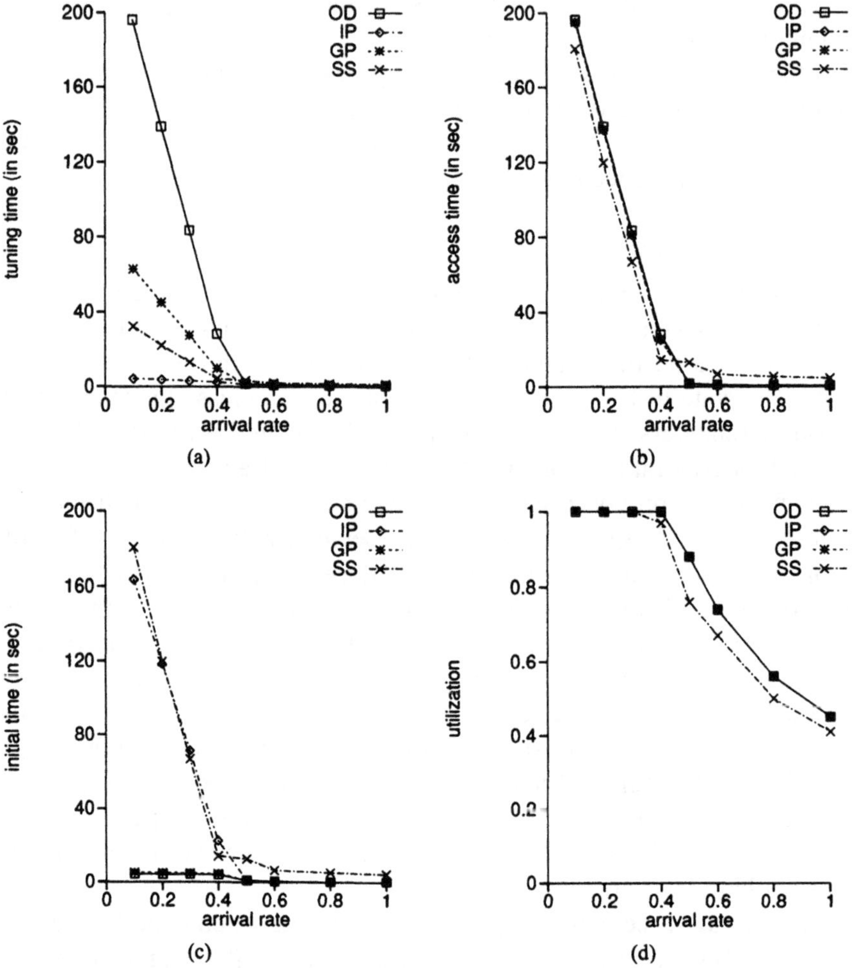

Figure 4.16. Comparison of pull-based selective tuning schemes.

It is also worth noting, from Figure 4.16(c), that OD performs best in the average initial access time. This is so since every client is listening to the broadcast, and objects that are broadcast in response to other queries may also be relevant to it. SS performs the worst in most cases for the same reason given above for its access time performance since the client can only receive its data objects, if any, when it is in active mode.

Finally, we note that the utilization of the bandwidth is the same for all schemes except SS which is slightly better (see Figure 4.16(d)). This also helps to explain why SS performs poorer than the other schemes under light load – it may doze off "unnecessarily" and fail to utilize the bandwidth effectively.

Notes

1 Most of the existing works focused on single object retrievals, and assume that the key value of the target object is provided in the query.

2 The scheme works with any kind of indexes. However, the access protocol may be different.

3 The original scheme broadcasts the index structure in preorder fashion only. In our variant, this is extended to the data file as well.

4 In fact, we only need h_{hash} as $h_{partition}$ is actually used in h_{hash}, i.e., $h_{hash} = f(h_{partition})$. But, for simplicity in presentation, we treat the two functions separately.

5 In [77], the term 'information frame' is used instead to refer to the logical unit of information broadcast on the air.

6 We note that the protocol in [77] is only for a 2-level signature approach.

Chapter 5

FAULT-TOLERANT BROADCAST ORGANIZATION

Data delivery in wireless communication environments is vulnerable to noise and signal distortion. Under the push-based model of information dissemination, if some of a mobile client's data are "corrupted" (and cannot be corrected), the client may have to wait for the next and even subsequent broadcast cycles to receive the desired data. This not only increases the access time of the data retrieval significantly but also the tuning time (or energy consumption) as well. Thus, the issues of retrieving data correctly and recovering from errors are important.

In this chapter, we shall examine the issue of broadcast organization in the context of unreliable wireless channels. In particular, we shall restrict our discussion to the cases when data are lost or corrupted. We assume that there is a mechanism for clients to detect that its data are corrupted. One standard technique is the cyclic redundancy code [116] which is low in overhead of redundant information and efficient in computation. Moreover, we shall focus on techniques that facilitate selective tuning to minimize energy consumption and access time. Traditional methods that multiplexed indexes with the data are not very helpful here because some of the index buckets along the path of the search process may be corrupted. In such cases, the search process may have to be restarted from scratch which can lead to high access and tuning time. We shall look at several fault tolerant indexing schemes that are variations of the tree-based, hash-based and flexible index schemes [110].

5.1 FAULT ON AIR

Wireless transmission is, unfortunately, error-prone. Many factors contribute to this. Signal may be lost due to external interference such as electromagnetic noises or obstruction by tall buildings and trees. The signal may also be weakened due to the clients' further distance from the server or the high speed at which the clients are moving. In addition, data may also be delayed when the clients move from a cell to another (since the data may not be broadcast in the same manner). Data may also be delayed due to congested network traffic.

Recall that in the push-based model of information dissemination, the server periodically broadcasts data to mobile clients. Each version of the broadcast (broadcast cycle) comprises both the data and the associated index information. Furthermore, the index and data objects are organized into buckets, where a bucket is the smallest unit of broadcast and access.

Data loss can occur at any bucket — index or data. With selective tuning, data buckets are examined only at two occasions — during the *initial probe* (when the client first tunes to the broadcast channel) and the *final probe* for the target data bucket. For the former, we need to probe repeatedly till a bucket that is error-free is obtained. For the latter, there is nothing to be done except for the client to wait for the next appearance of the data item in the next cycle. This may involve restarting the searching process (unless the cycle length is fixed and the clients know the cycle length).

If an access failure occurs while navigating through the sequence of index buckets, the problem is much harder to handle. This is because the sequence of access is critical for clients to continue the search process, and a failure will render the sequence out of order. Under existing selective tuning schemes, there is no mechanism for the clients to continue searching. Instead the clients have to *restart* the search process or *reprobe* for the erroneous index bucket (provided the cycle length is known and fixed). Both methods can lead to high access time — restarting does not guarantee that errors will not occur in the new sequence of probes and may miss the data if it is in the same segment as the erroneous bucket; reprobing implies waiting for the next broadcast cycle for the same bucket to be available. Furthermore, in the event that the sequence of index buckets is corrupted but the target data bucket is not, waiting for a complete broadcast cycle (which can be long) is certainly unnecessary and undesirable. Similarly, tuning time will also be higher when there is a failure since more index buckets will have to be downloaded and accessed.

Figure 5.1 illustrates a data broadcast. In the figure, the file has been split into k data segments, $D_1, D_2, \ldots, D_k$. Associated with each data segment is an index segment. Note that the index segment directs the search to data in other data segments, and provides direct offset to data in its associated data segment. As shown in the figure, a broadcast cycle comprises k index segments and k data segments. When the channel is error-free, the access time corresponds to the time from the moment the client tunes in to the time when the desired data is downloaded. However, if failure is not impossible, then the access time can be much larger. For example, suppose the client first tunes in at segment D_{k-1} of the $(j-1)$th cycle and its desired data is in segment D_2. Suppose also that the access sequence if the channel is reliable is: buckets in I_k, buckets in

I_1, buckets in I_2 and target data bucket in D_2. Suppose an error occurs when probing the sequence of buckets in I_2. Then, either restarting the search or reprobing will imply that the client will miss the target bucket in D_2 resulting in the access time being increased by at least one broadcast cycle. Now, if the target data bucket is not corrupted, then the client might have accessed the data in the same access time as an error-free channel.

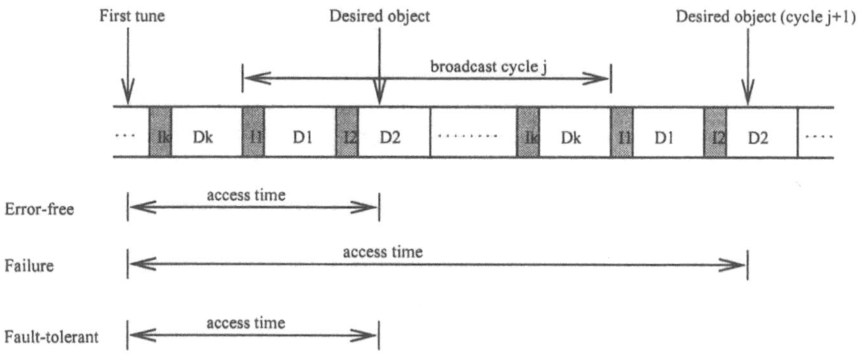

Figure 5.1. A data broadcast.

While tuning time is critical, the access time should not be excessively high. Instead, it is important to design access methods that can avoid unnecessary increase in the access time without incurring excessive tuning time. In other words, if the target data bucket is not corrupted, then it should be retrieved without having to wait for the next broadcast cycle, i.e., selective tuning mechanisms should be used to reduce the tuning time without significantly increasing the access time.

5.2 CHARACTERISTICS OF FAULT TOLERANT INDEX

Fault tolerant indexing schemes are typically based on *index replication*. A straightforward method is to duplicate each index bucket to reduce the bucket read error from p to p^2, where p is the failure probability of a channel. However, such a strategy will double the cycle length of an index tree. Instead, an effective scheme should have the following properties:

1. The degree of replication should be small. This is so since replication increases the broadcast cycle length and may have a negative effect if not properly controlled.

2. The phenomenon of reaccess from scratch should be minimized. This is important because the more data a client reads, the more prone the client is to error. In other words, the probability of an access failure can be reduced

by minimizing the amount of data read. For example, suppose the sequence of index bucket accesses is $a_1, a_2, \ldots, a_h$. Suppose an error occurs at a_k ($k < h$). Then, using a traditional method requires reaccessing $a_1, a_2, \ldots, a_k - 1$. Clearly, the next error can possibly occur at any of these ($k - 1$) accesses again. On the other hand, if we can avoid reaccessing $a_1, a_2, \ldots, a_k - 1$, we are certainly better off.

3. It would be a nice property that upon a failure, it is still possible that the client can restart the tuning process with the next index bucket within the same broadcast cycle (rather than waiting for the next version of the index). For example, consider a tree-based index with 5 levels. Suppose the failure occurs at level 3, then it would be better to allow the client to somehow continue from level 4 of the same tree rather than to wait for the next index tree to be broadcast (which may mean increasing the access time by at least one broadcast cycle).

5.3 INTER-INDEX SCHEMES

In *inter-index* schemes, upon an access failure, the search process continues from the next copy of the index. As such, these schemes do not have property 3. However, these schemes are still better off than traditional techniques. Essentially, inter-index schemes replicate the offset (pointers) to index nodes. Depending on the degree of replication, we can arrive at a family of such schemes. We shall focus on two tree-based methods proposed in [25].

The All-Children Replication Method (ACR)

The first method is called the *all-children replication scheme* (ACR). Under ACR, each index entry points to two occurrences of its descendant—one in the current broadcast and the other in the next broadcast. In other words, there is an additional "look-ahead" pointer for each entry to avoid reaccessing at the root of the next index tree; instead, the search can be continued from where the failure occurs in the next index. Figure 5.2 shows an example. Here, suppose the client succeeds at bucket $N_{i,j}$ but fails at its child node, say $N_{i+1,x}$. As bucket $N_{i,j}$ has for each of its children a spare pointer directed to its occurrence in the next index tree, the client can follow the spare pointer to the next occurrence of $N_{i+1,x}$ and continue the search process from that point on.

The client's protocol to access an object with key K is as follows:

1. Tune to the broadcast for an uncorrupted bucket. Get the offset to the next index bucket, and doze off.

2. Tune to the beginning of the designated bucket and examine the meta-data. If the bucket is corrupted, goto 1; otherwise (the bucket is error-free),

   - Keep track of the spare offset in the bucket.

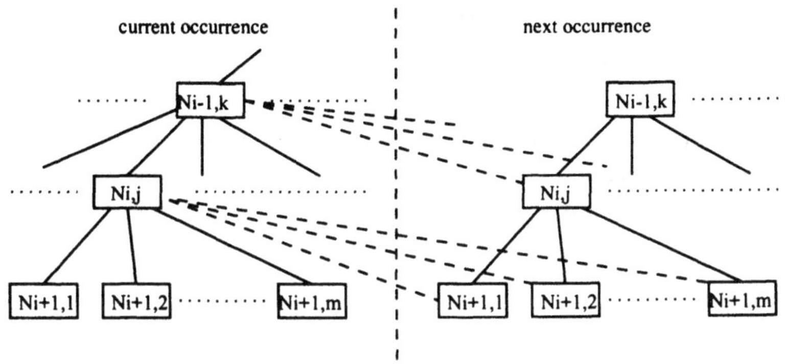

Figure 5.2. The all-children replication scheme.

- If the desired object has been missed, doze off till the beginning of the next cycle of the broadcast. Goto 3.
- If the desired object is not within the data segment covered by the index bucket, doze off to the next higher level index bucket. Goto 3.
- If the desired object is within the data segment covered by the index bucket, goto 3.

3. Probe the designated index bucket. If it is error free,

   - Keep the spare pointer, determine the offset to the next bucket along the path to the data bucket containing the target object. If the next bucket is an index bucket, doze off and goto 3, else (i.e., data bucket) doze off and goto 4.

   otherwise,

   - If there is a spare pointer, use it to determine the offset of the index bucket in the next version of the broadcast. Reset the spare pointer to NULL. Goto 3.
   - If there is no spare pointer, goto 1.

4. Tune in again when the bucket containing objects with key K is broadcast, and download the bucket (and all subsequent buckets as long as they contain objects with key K) if there is no error. If any bucket is corrupted and there is a spare pointer, goto 4, else goto 1.

Clearly, this strategy cannot survive two successive failures. Upon such failures, clients will have to restart its search process from scratch (i.e., from the root of the next index). Anyway, the probability of such situation is expected to be low — p^2, where p is the fault probability of one index bucket.

The Self Replication Scheme (SR)

The problem with ACR is that, if the number of children is large and the size of a link is not very small, the size of the total index tree may become very large. To cut down on the size of the tree, instead of replicating every child pointer, we can replicate node pointers. In other words, in each index bucket a single link is placed that points to the next occurrence of the node itself (see Figure 5.3).

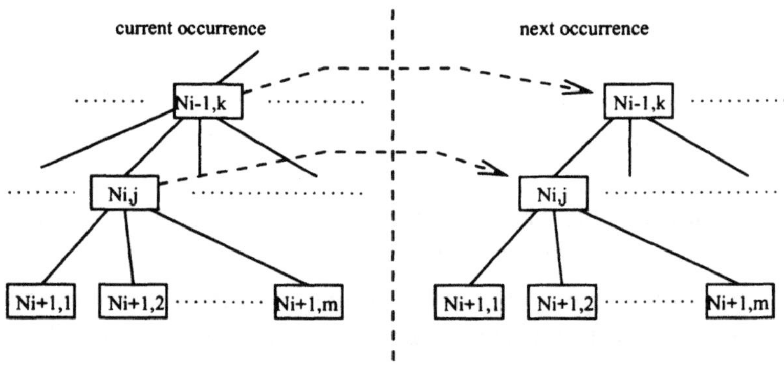

Figure 5.3. The self replication scheme.

In the *self-replication strategy* (SR), each non-leaf node will have one additional pointer that points to the same node in the next broadcast. Referring to Figure 5.3, when $N_{i,j}$ is successfully accessed, the path from the root to $N_{i,j}$ is "memorized". Upon failure at node $N_{i+1,x}$, as the client has succeeded at node $N_{i,j}$, therefore, it can use the spare link to find the next occurrence of $N_{i,j}$, and start the retune process from that point on. Compared to strategy ACR, SR has the disadvantage that, upon failure, the tuning process is forced one step back. Two successive failures will also invalidate all its earlier efforts. The access protocol for this scheme is essentially the same as that for the ASR scheme.

We note that both the ACR and SR schemes can be generalized to look ahead to k occurrences in order to reduce the fault probability to p^{k+1}. This will incur the overhead of larger index tree structure. Furthermore, we also note that if the file to be broadcast is static and will never be updated, then the schemes can be optimized. In such a context, there is no need to keep the offset for each object (or node) as the broadcast would be static and every cycle is essentially the same. Instead, only the broadcast cycle length is needed since every object is repeated in exactly one broadcast cycle away.

5.4 INTRA-INDEX SCHEMES

Both strategies ACR and SR have the same deficiency that, upon failure, they will direct the tuning process to the next index tree. This is undesirable since if the desired object is in the current segment, failure will increase the access time by another full cycle length. Thus, whenever possible, the reaccess process should start within the same broadcast cycle. These techniques, referred to as the *intra-index* schemes, try to continue the search process from the same broadcast cycle from which an access failure has occurred. We shall look at three such schemes studied in [110] which are respectively enhancements of the tree-based [60], hash-based [61] and flexible index schemes [61].

5.4.1 Tree-based Scheme

In the basic tree-based indexing scheme (also known as the *distributed indexing* scheme), a data file is associated with a B^+-tree index structure (see Figure 5.4). The broadcast organization is obtained by a preorder traversal of the tree. Furthermore, the first k levels of the index will be *partially* replicated in the broadcast, and the remaining levels will not to be replicated. The resultant broadcast organization is shown in Figure 5.5.

The major problem with the basic tree-based scheme is that there is no mechanism to handle access failures. Consider the example broadcast organization in Figure 5.5. Suppose a client tunes in at d_1 and is interested in accessing data object 40. If there is no access failures, the access path would have been: d_1, second a_1, b_2, c_5, d_{14} and data bucket containing object 40. However, if a failure occurs at b_2, then assuming there are no more failures, the access path becomes: d_1, second a_1, b_2 (error), c_4, third a_1, second I, first a_1, b_2, c_5, d_{14} and the target data bucket. The consequence is that the access time is increased by one broadcast cycle. This increase is unnecessary if the target data bucket is not corrupted at all.

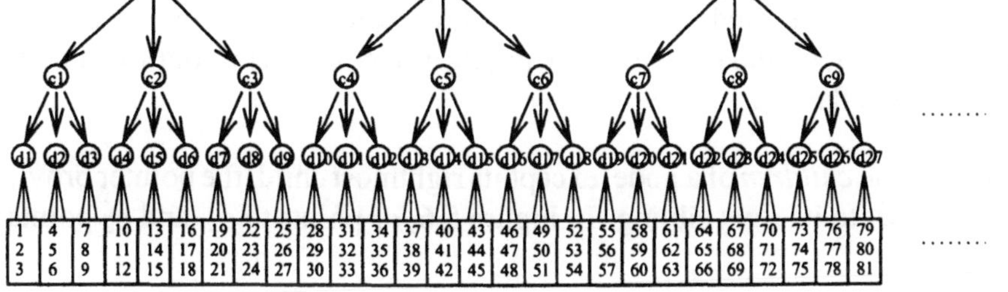

Figure 5.4. A partial data file and its index tree.

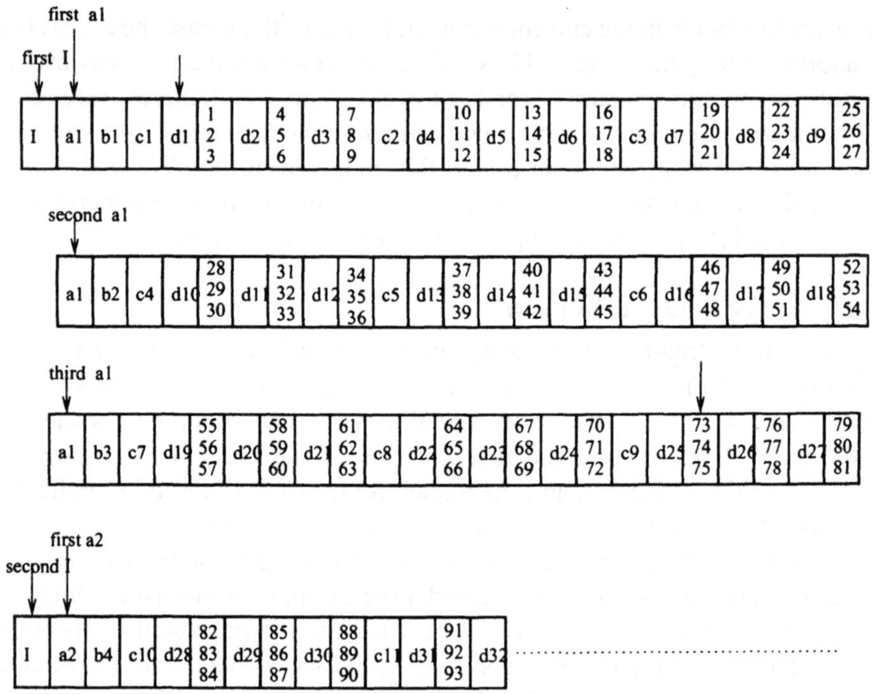

Figure 5.5. Partial data broadcast for the basic tree indexing scheme.

The basic scheme can be enhanced to be fault-tolerant by storing an additional (*maxvalue, pointer*) pair in each index node (except the root). The *maxvalue* of a node represents the largest data value that can be obtained for the subtree rooted at the node. Referring to Figure 5.6, *maxvalue* of nodes b_1 and c_2 are 27 and 18 respectively. For any node, its ancestor's *maxvalue* is greater than or equal to its *maxvalue*. Those ancestors whose *maxvalues* are the same are referred to as the E_ancestors of the node. We note that a rightmost child node of a node may have more than one E_ancestors. The one furthest away (closest to the root) is known as the F_ancestors. For example, d_9 has two E_ancestors, c_3 and b_1, and b_1 is its F_ancestor. Similarly, d_{27} has three E_ancestors, c_9, b_3 and a_1, with a_1 being its F_ancestor. The additional pointer comes in two flavors:

- For all the *children* of a node, except its rightmost child, the pointer provides the offset to its right sibling. In Figure 5.6, c_1 points to c_2, and c_2 points to c_3.

- For the rightmost child of a node, its pointer points to the right sibling of its F_ancestor. Consider Figure 5.6 again. d_3 points to c_2 since c_2 is the sibling of its F_ancestor c_1. Using the same logic, we note that d_{27} and c_9 both point to a_2.

Figure 5.6 shows the tree structure. Its broadcast organization turns out to be exactly the same as that in the basic scheme, i.e., as shown in Figure 5.5.

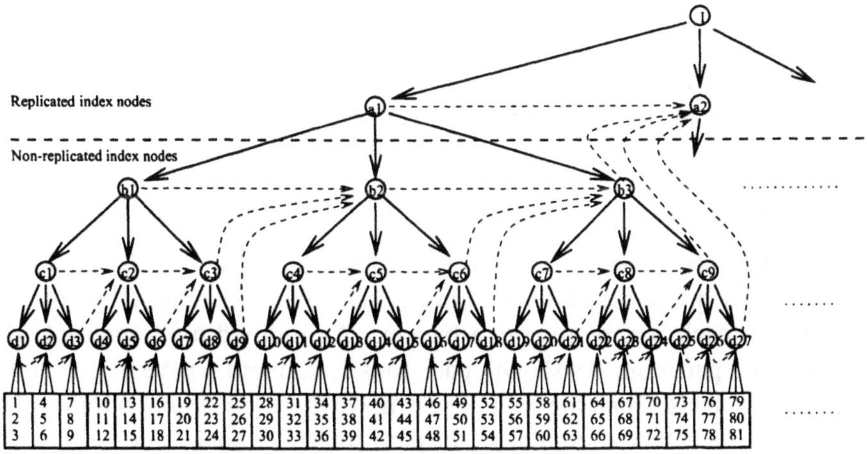

Figure 5.6. A fault tolerant index tree.

The revised selective tuning scheme works with the following access protocol (assuming the desired object is in the broadcast):

1. Tune to the current bucket and listen for an error-free bucket.

2. If the bucket is an index bucket, examine the entries to determine the offset of the next bucket that will eventually lead to the target data bucket. This may require examining the (*maxvalue, pointer*) pair. Switch to doze mode till the bucket is broadcast and goto 1.

3. If the bucket is a data bucket, examine the data bucket for the desired object. If the object is found, terminate; otherwise goto 1.

We use Figure 5.5 and Figure 5.6 for illustration. Suppose the desired data has key value of 54. Assume also that the initial probe occurs at d_1, then

- If there are no faulty accesses, the sequence of buckets to be accessed are: d_1, d_2, d_3, c_2, c_3, b_2, c_6, d_{18} and data bucket containing key 54.
- If bucket d_3 is faulty, then the sequence becomes: d_1, d_2, d_3, data bucket containing 7-9, c_2, c_3, b_2, c_6, d_{18} and the target data bucket.
- If bucket b_2 is faulty, then the sequence becomes: d_1, d_2, d_3, c_2, c_3, b_2, c_4, c_5, c_6, d_{18} and the target data bucket.

- If buckets b_2 and c_4 are faulty, then the sequence becomes: d_1, d_2, d_3, c_2, c_3, b_2, c_4, d_{10}, d_{11}, d_{12}, c_5, c_6, d_{18} and the target data bucket.
- If the target data bucket is corrupted, then the sequence becomes: d_1, d_2, d_3, c_2, c_3, b_2, c_6, d_{18}, the target data bucket, third a_1, second I, second a_1, b_2, c_6 and the target bucket again.

Note that the access time is the same for all the cases as long as the data page that contains the desired object is not corrupted. Thus, we can see that the proposed enhancements can maintain the expected access time while keeping the energy consumption low.

5.4.2 Hash-Based Scheme

In a hash-based scheme, data are hashed into a set of partitions that are broadcast as a sequence of buckets. Figure 5.7 shows an example of 4 partitions. In the figure, each box represents a bucket. An unfilled box represents the beginning of a partition. As shown, partitions 1, 2, 3 and 4 have 4, 6, 3 and 3 buckets respectively. For a file with N partitions, there are N *shift pointers* that indicate the offset to the beginning of the partitions. The buckets containing the shift pointers are referred to as the *hash buckets*. The interval between the shift pointers are called the *displacement*. In Figure 5.7, the shift pointers are placed at an interval of 3 buckets away. The first shift pointer has a value of +0 which means that partition 1 begins at the same bucket where the first shift pointer is stored. On the other hand, the fourth shift pointer is +4 which indicates that partition 4 is 4 buckets away.

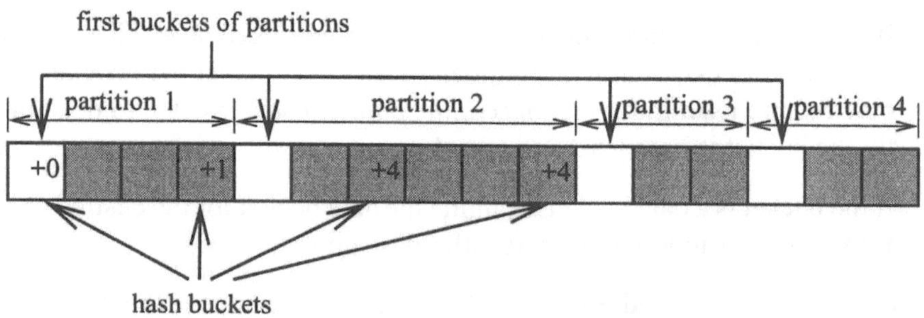

Figure 5.7. A broadcast with 4 partitions.

The basic hash-based scheme also *collapses* when the channel is error-prone. Consider the example shown in Figure 5.7. If the bucket containing the shift pointer is corrupted, then the client can no longer obtain the starting address of the partition, i.e., the client may have to redo its search process from scratch leading to an increase in the access time by at least one broadcast cycle. This is undesirable especially when the target partition may be far away from the hash bucket (e.g., partitions 3 and 4 in Figure 5.7) and the target partition is not corrupted.

Consider a n-partition file. Let P_i denote the ith partition, and h_i denote the ith hash bucket. To reduce the number of unnecessary restarts in an error-prone channel, the hash-based scheme can be modified as follows (see Figure 5.8):

- Control information can also be stored in some of the buckets between hash buckets. Consider two consecutive hash buckets, h_i and h_{i+1}. There are two possible scenarios:
  - The first bucket of partition P_i is broadcast before bucket h_{i+1}. For example, in Figure 5.8, partition 1 appears before bucket h_2, and partition 2 begins before h_3. Let the offset at h_i be j. Then, we can include a shift pointer in j buckets following h_i where bucket $(h_i + k)$ has shift pointer value of $(j - k)$, for $1 \leq k \leq j$. Note that bucket $h_i + j$ has a shift pointer of 0 and is in fact the beginning of the target partition. Referring to Figure 5.8 again. Since bucket h_2 has a value of +1, we can store a shift pointer in bucket $h_2 + 1$.
  - The first bucket of partition P_i comes at or after bucket h_{i+1} (except for h_n). For example, in Figure 5.8, partition 3 comes after h_4. Let the offset at h_i be j. Then, we can include a shift pointer in all the buckets between h_i and h_{i+1} (except for h_n). The pointer value for bucket $h_i + k$ is k less than that in h_i, for $1 \leq k \leq h_{i+1} - h_i$. Using the same example, we have included in buckets $h_3 + 1$ and $h_3 + 2$ shift pointer values +3 and +2 respectively. For h_n, we can include a shift pointer in all the buckets following h_n until we reach the beginning of partition n. As shown in Figure 5.8, partition 4 has shift pointers at h_4, $h_4 + 1$, $h_4 + 2$, $h_4 + 3$ and $h_4 + 4$.

  Clearly, with this simple enhancement, unless all buckets containing the shift pointers to the target partition are corrupted, the need to restart will be very small.

- The objects of a partition are sorted. In this way, the client can terminate the search process once the objects are found. More importantly, if some of the data buckets of a partition are corrupted, the client may not need to restart as long as its desired data have not been missed.

- Each bucket also contains the partition id. Note that the partition id can take on the same value as the hash bucket id.

Let the search key be K, and the object be in partition with id *pid*. For simplicity, we assume that the partition id is the same as the hash bucket id. The revised access protocol becomes as follows:

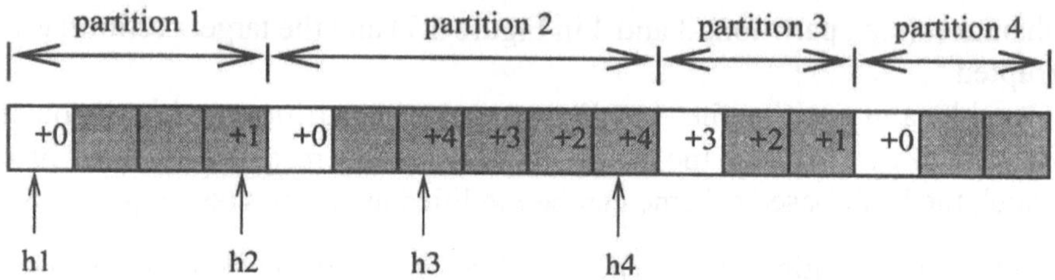

Figure 5.8. A fault tolerant hash-based scheme of Figure 5.7.

1. Tune to the current bucket, listen for an error-free bucket, and determine the hash bucket for K. Let the target hash bucket be h_i, and the bucket id of this current bucket be bid.

2. Determine N_{doze}, the number of buckets to doze off before slot h_i is broadcast. Note that there are three cases to consider in determining the number of buckets to doze off:

   - *The target hash bucket will be broadcast in this cycle soon.* If $bid \leq h_i$, then $N_{doze} = h_i - bid$. For example, if $bid = 4$ and h_4 is the target hash bucket, then $N_{doze} = 9 - 4 = 5$.
   - *The target hash bucket in this cycle has been missed.* In this case, there are two possible scenarios:
     - *The hash bucket and the target partition have been missed.* If $bid > h_i$ AND $pid > h_i$, then N_{doze} is given by the sum of the offset to the beginning of the next version and h_i. For example, if $bid = 10$ and h_2 is the target hash bucket, then $N_{doze} = 6 + 3 = 9$.
     - *The hash bucket has been missed but the target partition has not.* If $bid > h_i$ AND $pid \leq h_i$, then the bucket is either part of the target partition or has yet to be broadcast in the cycle. In this case, $N_{doze} = 0$. For example, if h_2 is the target hash bucket and $bid = 5$, then $N_{doze} = 0$. As another example, if h_3 is the target hash bucket and $bid = 8$, then $N_{doze} = 0$.

3. If $N_{doze} = 0$, then

   - If $bid < h_{i+1}$ AND $pid < h_i$, obtain the *shift* value. Note that this shift value is the "spare" *shift pointer* provided by the enhanced scheme. Switch to doze mode until *shift* number of buckets. Goto 1.
   - If $bid \geq h_{i+1}$ AND $pid < h_i$, then though the target partition is still not broadcast, there is also no shift pointer to facilitate doze off. In this case, either all the buckets are scanned until the target partition is found or the situation is handled as if the partition has been missed. For

the former, the tuning cost may be high especially if the target partition is very far away (when the minimum displacement is small). As a compromise, if $h_i - pid < \tau$ (τ serves a threshold), all the buckets are scanned, i.e., goto 1. Otherwise, N_{doze} is recomputed based on the case when the target partition has been missed.

- Otherwise (i.e. $pid = h_i$), examine the objects in the bucket. Since the objects are sorted, even if the client tunes in mid-way through the partition, the client can tell whether the desired object has been missed.
  - If the target object is found, terminate.
  - If the first object has key $> K$, then the desired object has been missed and the number of buckets to doze off is given by the sum of the offset to the beginning of the next version and h_i. Goto 1.
  - If all the object keys $< K$ (i.e., data not found in current data bucket), goto 1.

4. If $N_{doze} > 0$, doze off for N_{doze} buckets (i.e., until slot $h(K)$ is broadcast). Goto 1.

We use Figure 5.8 to illustrate the access protocol. Suppose the desired object is in bucket 2 of partition 3. Let us use P_{ij} to denote the bucket j of partition i. Suppose the initial probe occurs at P_{11}. Then

- If there are no faulty accesses, the sequence of buckets to be accessed are: P_{11}, P_{23}, P_{31} and P_{32} (the target bucket).
- If bucket P_{23} is faulty, then the sequence becomes: P_{11}, P_{23}, P_{24}, P_{31} and P_{32} (the target bucket).
- If buckets P_{23} and P_{31} are faulty, then the sequence becomes: P_{11}, P_{23}, P_{24}, P_{31} and P_{32} (the target bucket).

Note that the access time is the same for all the cases as long as the bucket that contains the desired object is not corrupted.

5.4.3 Flexible Indexing Scheme

The flexible indexing scheme splits a sorted list of objects into several equal-sized segments, and provides indexes to navigate through the segments. This is achieved as follows. At the beginning of each segment, there is a pair of indexes: a *global index* and a *local index*. The global index is used to determine the segment which an object may be found, while the local index provides the offset to the portion within the segment where the object may be found.

Figure 5.9 shows a sample flexible index scheme. Here, the file is divided into 8 segments of 4 objects each (except for segment 8). Consider segment 2. There are 3 entries in the global index. The first entry, (12, 27), implies that any object that is less than 12 can be found after the 27th object, i.e., in the beginning of the next version of the file. The second entry, (54, 13), means that an object larger than or equal to 54 can be found after the 13th object, i.e., in segment 5. Similarly, the third entry (42,5) indicates that an object between 42 and 54 may be found 5 entries away, i.e., two segments away. Implicitly, objects that may be found in the segment follows the local index. The local index has 3 entries. Consider the local index of segment 2. The entry (29, 4) means that an object with value 28 and above may be found after 4 objects away. The entry (22, 3) refers to an object with value above 22 but less than 29 can possibly be found 3 objects away. Likewise, the entry (20, 2) points to where objects between 20 and 22 can be found. Implicitly, any object less than value of 22 can be found at the end of the local index.

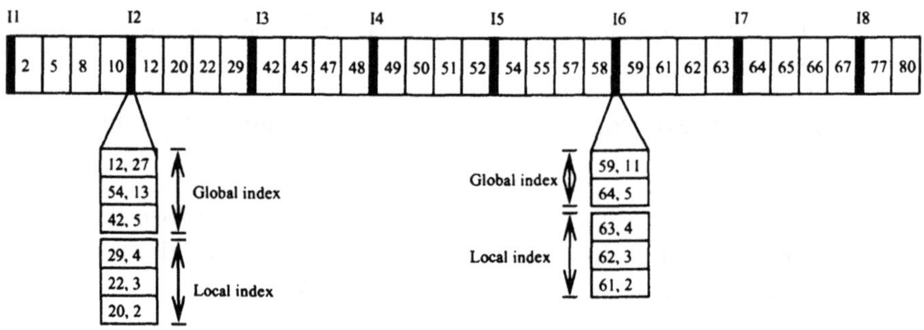

Figure 5.9. Flexible index scheme.

As in the other schemes, the flexible index scheme is effective when the channel is reliable. However, when the global index is corrupted, the client will have to restart with the next segment's global index in order to be able to proceed on with the search.

It is interesting to note that the flexible scheme requires minimal modification. For each data bucket, instead of an offset to the next segment, a (*maxkey*, *offset*) pair is maintained. While the offset remains the same, the *maxkey* indicates the largest key value within the segment. Given a search key, K, the access protocol is also revised to become as follows:

1. Tune into the channel and listen for an error-free bucket.

2. If bucket is a data bucket, we have two scenarios. Suppose x is the first value in bucket.

   - If K >*maxvalue* or $K < x$, then obtain the offset from bucket, doze off till the next segment is broadcast. Goto 1.

   - If $x < K$ <*maxvalue*, examine all objects in bucket. If found, terminate; otherwise, goto 1.

3. If bucket is an index bucket
   - If it is a global index bucket, examine its entries.
     - If the target object falls into a segment whose offset is located in this index bucket, get the offset, doze off for the appropriate amount of time and goto 1.
     - Otherwise (i.e., there are several global index buckets and the current bucket does not contain the desired offset) goto 1 (i.e., read the next bucket).
   - If it is a local index bucket, examine its entries.
     - If the object can be found in some sections in the current local index bucket, obtain the offset to the respective section, switch to doze mode for amount of time determined by offset, goto 1.
     - Otherwise, goto 1.

To illustrate, let's use Figure 5.9 as our example. Suppose each bucket contains only one data object, and each global index is contained in a bucket, and each local index is contained in a bucket. Suppose the search key is 61. Then,

- If the client tunes in at object 59 and there is no fault, the access path will be: 59 and 61.
- If the client tunes in at object 10 and there is no fault, the access path will be: 10, I2's global index, I5's global index, I6's global index, I6's local index, and 61.
- If the client tunes in at object 10, and an error occurs at I2's global index, then the access path becomes: 10, I2's global index, I2's local index, 12, I3's global index, I6's global index, I6's local index, and 61.
- If the client tunes in at object 10, and errors occur at I2's global and local indexes, then the access path becomes: 10, I2's global index, I2's local index, 12, I3's global index, I6's global index, I6's local index, and 61.

Eng and Tan [41] further extended the fault tolerant flexible scheme to deal with non-flat broadcast in unreliable wireless environments.

5.4.4 Discussion

A simulation study was conducted to investigate the effectiveness of the fault tolerant tree-based, hash-based and flexible index schemes. The results showed that the various fault tolerant schemes can outperform their counterparts in terms of access time without sacrificing much on the tuning time. Here, we compare the relative performance of the three schemes. The tree-based, hash-based and flexible-based schemes are denoted as Hash, Tree and Flexible respectively. Figure 5.10 shows the results as the error rate increases. First, we observe that the tree-based scheme performs the worst in terms of access time. This is because of its broadcast cycle which is the longest among the three schemes. The hash-based scheme is superior when the error rate is high and the flexible-based method performs the best at lower error rate. Two factors contributed to this result: the cycle length of the broadcast (flexible-based method has a longer broadcast cycle than the hash-based scheme) and the minimum displacement of the hash-based scheme. In general, the hash-based scheme performs poorer because of clients who miss the hash buckets in the initial probe. As the error rate increases, the contribution of the longer cycle length of the flexible scheme to the access time becomes significant, leading to its poorer performance at high error rate.

In terms of tuning time, we notice that the tree-based scheme is superior under low error rate, but degrades speedily as the error rate increases. The hash-based scheme performs poorly at low error rate by virtue of its need to examine records of the target partitions. The flexible scheme performs slightly worse than the tree-based scheme under low error rate, and is the most energy efficient under high error rate.

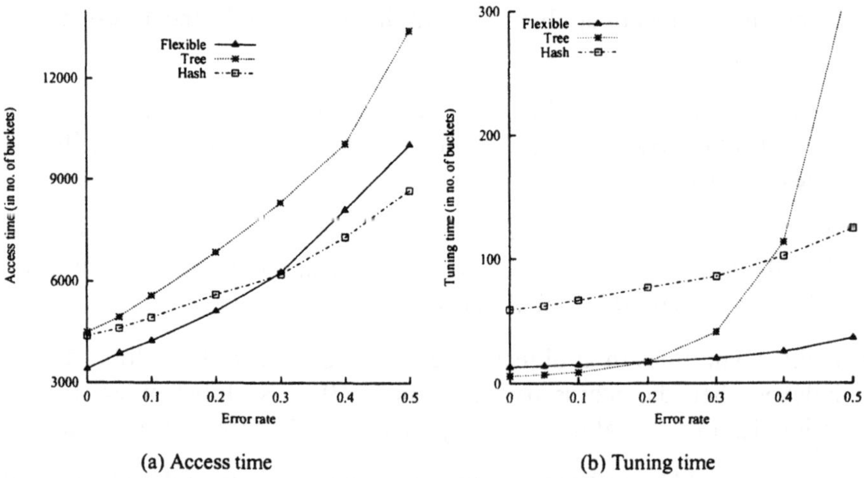

Figure 5.10. Comparison of the three fault-tolerant schemes.

Chapter 6

CACHE INVALIDATION MECHANISMS

In wireless computing environments, mobile clients can cache frequently accessed data objects. They can do this in a disk or any storage system that survives power disconnections, such as flash memories. In fact, caching has been considered to be a very effective mechanism for wireless computing. First, clients' needs can be met with the cached objects. This implies short access time and, more importantly, that contention on the narrow wireless bandwidth can be minimized. Second, clients do not need to expend energy to transmit or receive data objects that are already cached. This can extend the effective battery lifespan. Finally, minimizing transmission can result in cost benefits as well, especially when one is billed on a pay-per-packet basis (e.g., Cellular Digital Packet data).

However, for caching to be effective, the cache content must be consistent with those objects stored in the server. This is, unfortunately, difficult to enforce in a mobile environment due to the frequent disconnection and mobility of clients. Clients who resume connection (after disconnecting) no longer know whether their cached content is still valid.

Traditional techniques require either the server to transmit invalidation messages to the clients every time an object is updated (e.g., the Andrew File System [93] and Coda [94]) or the client to query the server to verify the validity of the cache content (e.g., the Network File System under the client-server architecture [90]). Both approaches, unfortunately, are not adequate for wireless environments. In the first approach, which has also been referred to as *stateful-based* approach, the server must keep track of the clients' cache content and locate the appropriate clients whenever an object is updated. Moreover, disconnected clients' cache content are automatically invalidated since they cannot be reached and would have missed the invalidation messages of updated objects. In addition, even if a client is not about to use a particular cached object, it gets notified about its invalid status, which is a potential waste of bandwidth. In the second approach, the client must send a message every time they want to use their cache. This is both wasteful of bandwidth and energy of the portable device.

In this chapter, we shall examine novel cache invalidation strategies that consider the factors of client's frequent disconnection and limited battery power supply. Moreover, we shall focus on the class of *stateless-based* techniques, where the server is not aware of the state of the clients' cache. We shall first present a framework for designing stateless cache invalidation schemes, from which a large class of cache invalidation schemes can be constructed. Second, we shall survey some of the representative techniques obtained from the framework that have been proposed in the literature.

6.1 A TAXONOMY OF CACHE INVALIDATION SCHEMES

In stateless-based cache invalidation schemes, the server has no information about which clients are currently under its cell and what the contents and "ages" of their cached objects are. To facilitate cache invalidation, the server broadcasts *invalidation reports* containing information on objects that are most recently updated, and the clients will listen for and use the reports to invalidate their caches. To our knowledge, all reported cache invalidation schemes in the literature are stateless-based [15, 26, 28, 66, 107, 123].

For the stateless-based approaches, we can further categorize invalidation methods into either *asynchronous* or *synchronous* methods [15]. Under the asynchronous method, once an object is updated, the server broadcasts its updated value immediately. The asynchronous method is effective for connected clients, and allows them to be notified immediately of updates. However, for a client who reconnects after a period of disconnection, the client has no idea of what has been updated and so the entirety of its cache content has to be invalidated. To salvage the cache content, Barbara and Imielinski [15] have proposed that an invalidation report can be piggybacked with each invalidation notice. In this case, upon reconnection, clients will have to wait for the first asynchronous invalidation report. However, since the report is sent asynchronously, there is no guarantee on how long the client must wait.

On the contrary, the synchronous method is based on periodic broadcasting of invalidation reports. The server keeps track of the objects that are recently updated, and broadcasts these information to clients periodically. Clients must first receive the invalidation report to conclude whether its cache is valid or not. If the cache is valid for the query, it can be used to answer the query; otherwise, the query will have to be submitted to the server. We note that waiting for the report adds some latency to query processing. But, because of its periodic broadcast nature, synchronous methods provide a bound on the waiting time of the next report.

In this section, we focus on *synchronous stateless-based* approaches. From existing work, we have identified some common issues that have been addressed in designing cache invalidation schemes. These are the *content of the report*, the *invalidation process*, and the information (i.e., *log*) that the server must maintain. Figure 6.1 summarizes these issues and the possible solutions.

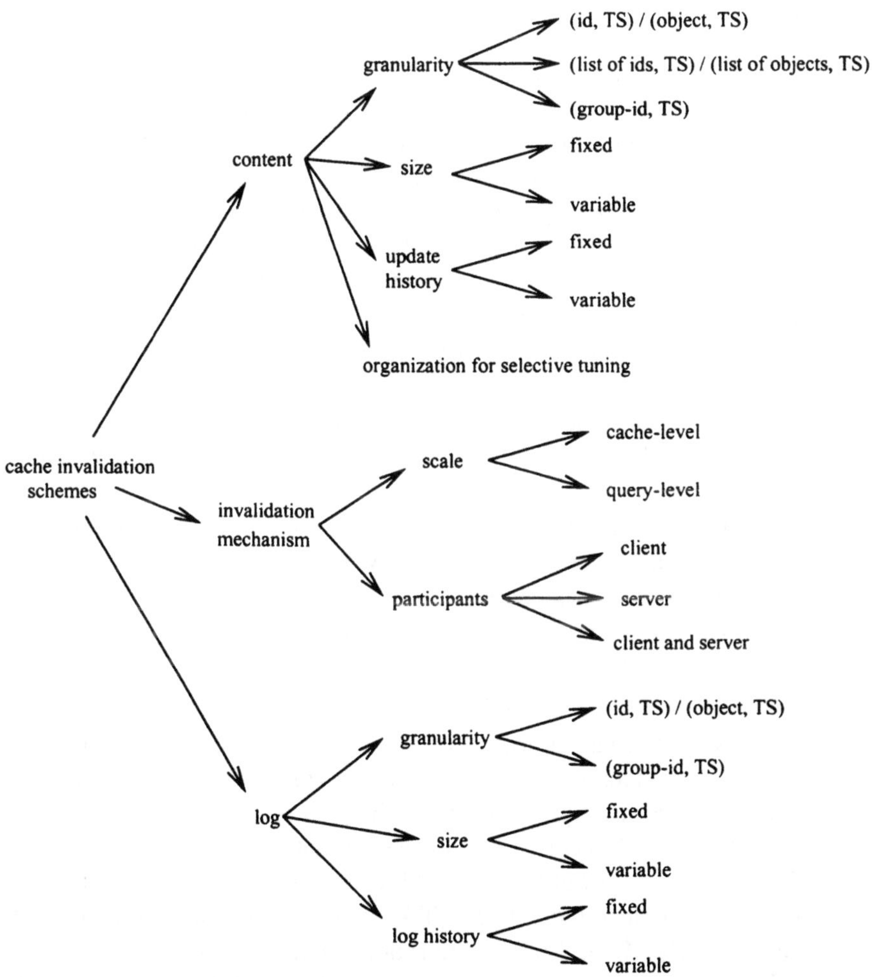

Figure 6.1. Issues and solutions in designing cache invalidation schemes.

6.1.1 On the Content of the Invalidation Report

Ideally, the server should keep track of all updates and broadcast them to the mobile clients. But, this is costly and impractical in view of the limited bandwidth and short battery life of mobile clients. Instead, the server maintains a short (of reasonable length) *history* of updates and broadcasts an invalidation report that reflects the most recent changes. Based on the report, clients can, to some extent, determine the validity of their cache content. Several issues need to be addressed with regards to the content of the report.

Granularity

The *granularity* of the report refers to the level of details of information each record of the report captures. A record in the report can be an (id,TS) pair, where id is the identifier of an object that has been updated and TS is the timestamp at which this object was updated. Alternatively, the report can reflect the full detail of an updated object at time TS, i.e., the record is the pair (object,TS). The former is commonly known as *update invalidation* as clients can only invalidate their cache content. The latter, on the other hand, allows clients to immediately update their invalid copy with the object that is broadcast. It is thus referred to as an *update propagation* mechanism. Clearly, there is a tradeoff between the two mechanisms. Under update propagation, when the disconnection time is short, clients can update its cache immediately. Under update invalidation, clients must still submit requests to retrieve the updated records even if the disconnection time is short. However, under update propagation, since the entire record is broadcast, the report is much larger and can take up a significant portion of the downlink channel capacity which is a scarce resource in wireless environment. Moreover, given the same report size, update invalidation can afford to reflect a longer history of updates. Thus, existing works are largely based on update invalidation [15, 26, 28, 107, 123]. Works on update propagation can be found in [28].

Instead of associating each object (or its id) with a timestamp in the report, an alternative is to associate a list of ids/objects with a timestamp. The report would then comprise multiple (list of ids, TS) pairs (or (list of objects, TS) pairs). Each (list of ids/objects, TS) pair means that the list of ids/objects have been updated since TS. For example, if objects O_1, O_2 and O_3 were updated at TS_1, TS_2, and TS_3 respectively where ($TS_1 < TS_2 < TS_3$), then we may have the record ([O_1, O_2, O_3], TS_1) in the report. This approach can reduce the number of timestamps needed. However, it may lead to *false* invalidation. For example, if a client accesses the three objects at time TS_d, $TS_2 < TS_d < TS_3$, and reconnects at time TS_r, $TS_r > TS_3$, then even though O_1 and O_2 are still valid, they would be invalidated. This approach has been studied in [66]. To cut down on the bandwidth requirement, the list of objects can be further *compressed* into a *signature*, i.e., (signature, TS) pairs are broadcast instead. This, however, will further increase the number of false invalidation as some valid objects' signatures may match the signatures of updated objects.

Another alternative that can minimize the usage of bandwidth for the invalidation report is to disseminate only *group-based* information. In other words, the database objects are organized into groups. The invalidation report contains (group-id, TS) pairs to reflect the most recent timestamp TS at which an object in the group (identified by group-id) has been updated. Using such a report, the client will invalidate all objects in the group if the group has recently been updated. The main problem with the group-based approach is that of false invalidation, i.e., objects belonging to a group that are valid may still have to be invalidated. Group-based schemes have been studied in [26, 107, 123].

Size and Update History

The size of the invalidation report can be fixed or varied. The update history refers to the history of the updates that are reflected in the report, and can be fixed or varied too. These two factors are interrelated in the sense that one typically affects the other. For example, the report size can be fixed by the amount of bandwidth allocated for the report. It can also be fixed by the number of objects/groups to be included in the report. Obviously, under these cases, the update history cannot be predetermined (i.e., it has to be variable) since the number of updates varies over a fixed period of time. On the other hand, report size can vary from broadcast to broadcast by fixing the update history to be reflected. As an example, an invalidation report broadcast at time T may reflect all updates during the interval [T-wL, T] for some $w > 0$ and L; w is referred to as the broadcast window and L is the fixed interval at which the report is periodically broadcast. Works that employ fixed report size with variable update history have been reported in [26, 66], while works on variable report sizes that fix the update histories have appeared in [15, 26, 107, 123].

Organization for Selective Tuning

To conserve energy, it may be necessary to organize the invalidation report to facilitate selective tuning. This can be done by interleaving the content of the report with "indexes" that can provide "direct" access to targeted portion of the report. Thus, only the desired portion of the report need to be examined. Organization of the report to facilitate tuning has been explored in [26].

6.1.2 On Invalidation Mechanism

There are two issues to address here. The first concerns the scale of the invalidation, whether it is cache-level or query-level. The second concerns the participants that are involved, whether the invalidation is performed by the *client only*, by the *server only* or a collaboration between the two.

Scale

When a client receives an invalidation report, it can invalidate its cache content in two ways. First, it can perform *cache-level* invalidation, i.e., cache validity is performed for all objects cached. This requires scanning a large portion of the invalidation report, if not the entirety of the report. As a result, it is power inefficient and not particularly suited for selective tuning (unless the number of cached objects is very small). On the other hand, the client can perform *query-level* invalidation, where validation is performed only on the objects queried. This reduces the number of objects to be invalidated and hence reduces the energy consumption. Moreover, since only a few objects are typically queried, the invalidation report can be organized for selective tuning. Examples of query-level invalidation schemes are SCI and OCI [26], while GCORE [123], Broadcasting Timestamps [15], Bit-sequences [66] and BGI [107] are cache-level oriented.

Participants

Invalidating the cache content can be performed by the client *alone*. This requires that the client bases its invalidation purely on the invalidation reports. Thus, the effectiveness of such approaches are dependent on the content of the report. On the other extreme, we can allow the server to perform the invalidation *alone*. This, however, will require the client to inform the server about its cache content which can be costly since transmitting these information consumes energy and bandwidth. Finally, the client and server can collaborate to identify the cache content that should be invalidated: the client uses the invalidation report to invalidate its cache content; for those that remain uncertain, the client submits their information to the server for invalidation. Broadcasting Timestamps [15], BGI [107], and SCI [26] are examples of schemes that employ client-only invalidation, while GCORE [123] adopts a client-server collaboration approach. The server-only approach was discussed in [106].

6.1.3 On Update Log Structure

Another important dimension in the design of a cache invalidation schemes concerns the information (update logs) maintained at the server to reflect the updates on the database.

The update logs may contain update information of each individual object or its identifier. For the former, the log record is of the form (object, TS) to reflect that object has been updated at timestamp TS. For the latter, the server only needs to maintain (id, TS) pairs, each of which indicates that the object with identifier id was updated at TS.

Alternatively, each log record may reflect updates on a collection of objects. In this *group*-based approach, objects are organized into groups, and the log record reflects the latest update to the group, i.e., each log record is of the form (group-id, TS) where TS is the most recent timestamp that an object in group identified by group-id has been updated.

The second issue concerns the size and log history (duration that the update logs should be maintained) which are interrelated. The size can be fixed by restricting updates to be maintained for a fixed number of objects. In this case, the log history changes depending on the updates. This approach has been adopted in Bit-Sequences [66] where updates to half the database are maintained.

On the other hand, variable sized logs can be maintained by fixing the log history to a fixed interval, say [T-WL, T] for some $W > 0$; W is the update log window, L is the fixed interval of broadcasting the invalidation report, and T is the timestamp at which the report is broadcast. In other words, only updates in the interval [T-WL, T] are maintained. Clearly, keeping all updates is ideal but can be costly in terms of storage, whereas keeping only very recent updates may not be effective as it may lead to false invalidation. This approach has been adopted in GCORE [123], Broadcasting Timestamps [15], BGI [107], and SCI [26].

6.1.4 Summary of Existing Cache Invalidation Schemes

Based on the above discussion, we can derive a large number of cache invalidation schemes from the product of the various alternatives in addressing each issue. It should be pointed out that some of the options may not be practical/feasible. For example, it is not possible to have a fixed size report that reflects all updates in a fixed update history [T-wL, T]. On the other hand, we can design techniques that combine multiple solutions. For example, there exists techniques (see Table 6.1) that reflect both recent updates for individual items, as well as group updates. Table 6.1 summarizes some of the existing work based on the dimensions discussed.

As can be seen from the table, most of the existing techniques are based on update invalidation rather than update propagation. This is to conserve the limited wireless bandwidth. Furthermore, except for the work in [26] (i.e., OCI and SCI), most of the work did not address the issue of selective tuning. In addition, OCI and SCI are the only query-level schemes. We also note that very few work (e.g., GCORE and UI-CS) adopt a client-server collaborative effort for invalidation. Such techniques incur the additional uplink overhead, which can be costly in terms of energy consumption (since transmitting data consumes more energy than receiving data). Most of the recent techniques

Table 6.1. Summary of Cache Coherency Schemes.

Algorithm	*Content*	*Invalidation Mechanism*	*Log Structure*
Broadcasting Timestamps [15]	(id, TS), size depends on number of updates in fixed update history [T-wL, T], no selective tuning	cache-level, client oriented	(id, TS), log history ([T-wL, T])
Simple Group [123]	same as Broadcasting Timestamps	cache-level, client + server	(id, TS)+(group, TS), log history ([T-wL, T])
GCORE [123]	same as Broadcasting Timestamps	cache-level, client + server	(id, TS)+(group, TS), log history ([T-WL, T])
BGI [107]	(id, TS)+(group, TS), variable size, fixed update history [T-wL,T] and fixed group history [T-WL,T] ($W > w$), no selective tuning	cache-level, client	(id, TS)+(group, TS), log history ([T-WL, T])
OCI [26]	(group, TS), fixed group size, variable update group history, selective tuning on groups	query-level, client	(group, TS), log history (depends on updates on groups)
SCI [26]	(id, TS)+(group, TS) variable size, fixed update history [T-wL,T] and fixed group size for updates in [T-WL,T], selective tuning	query-level, client	(id, TS)+(group, TS), log history ([T-WL, T])
Bit-Sequences [66]	(list of ids, TS), fixed size (database size), variable update history, no selective tuning	cache-level, client oriented	(id, TS), log history (variable (intervals, depends on database size)
UP-C [28]	(object, TS), variable size for updates in [T-wL,T], no selective tuning	cache-level, client oriented	(object, TS), log history ([T-WL,T], $W > w$)
UI-CS [28]	(id, TS), variable size for updates in [T-wL,T], no selective tuning	cache-level, client + server	(id, TS), log history ([T-WL,T], $W > w$)

also maintain a longer log history than the broadcast history (i.e., $W > w$) in order to salvage more of the cache content. Finally, we would like to point out that the work in [28] dealt with relational operations, while the rest focused on object retrievals.

6.2 CACHE INVALIDATION SCHEMES

In this section, we examine several synchronous cache invalidation schemes designed under the framework proposed in the last section. All the schemes assumed that the server broadcasts an invalidation report every L time units. This effectively divides the time into intervals. Moreover, clients must wait for the next invalidation report before answering a query. Waiting for the next invalidation report guarantees that all the objects are up-to-date as of the starting time of the report. This also means that the client has to keep a list of objects queried during an interval and answers them after receiving the next report. The consequences of this way of operation is two fold [15]: two or more queries on the same object posed during an interval will all be answered at the same time in the next interval; and the answer to a query will reflect any updates to the object during the interval in which the query was posed, and this is the case even if they query predates the update during the interval.

Throughout this section, we shall assume the server contains DB objects. The server also maintains an update history of objects during the last WL time units, where W is the update log window. The client maintains a cache of frequently accessed objects. Each cached object O has identifier o_{id} and is also associated with a timestamp t^c_{id} that reflects the last known update time of the object. The client also keeps the timestamp T^c that indicates the last time it received an invalidation report. The client also maintains a list of query objects, Q, in the interval $[T - L, T]$, where T is the time at which the next report will be broadcast. We will also use the following running example to illustrate the various schemes.

EXAMPLE 8 *We shall use a database of 16 objects as our running example. Figure 6.2 shows the database objects, and the timestamp at which each object has been updated. For example, object 5 was updated at time 22, and object 12 was updated at time 30. This example will be used to illustrate how the various cache invalidation schemes studied in this paper (in)validate the content of the client's cache.* □

6.2.1 Broadcast Timestamp

The *Broadcast Timestamp* strategy [15] is one of the first cache invalidation algorithms proposed, and has been widely used as a basis for comparison in many techniques that were subsequently proposed [26, 123]. It is based on the following parameters:

Database objects and update timestamps

ID	1	2	3	4	5	6	7	8	9	10	11	12	13	14	15	16
TS	24	16	10	6	22	18	26	32	2	20	14	30	8	4	12	28

Figure 6.2. A running example.

Content. The report consists of a list of (o_{id},t_{id}) pairs. The size of the report varies as it is dependent on a fixed broadcast window of wL time units. There is no selective tuning.

Invalidation Mechanisms. The invalidation is performed at the cache-level, and by the client only.

Log. The server maintains logs for objects. The size of the log varies though the log history is fixed at [$T - wL$,T] for objects, where T is the current time.

The basic idea is for the server to maintain the updates in the last wL time units[1], i.e., the list of (o_{id}, t_{id}) pairs where o_{id} is the identifier of an object updated in the interval $[T - wL,\ T]$, and t_{id} is the timestamp of the latest change for the object, $t_{id} > T - wL$. The server broadcasts an *object invalidation report* (OIR) periodically at times $T_i = iL$, where $i > 0$. The content of the report is exactly the update information maintained by the server.

Suppose the next OIR will be broadcast at time T. At T, the client listens to the report and uses it to (in)validate its cache content: if the difference between the current report timestamp (i.e., T) and the timestamp of the last report that the client has received (i.e., T^c) is larger than wL, the entire cache is discarded; otherwise, for each object cached, the client either removes it from the cache (when its id appears in the invalidation report and its timestamp in the report is larger than that of the cached copy) or updates its timestamp to the timestamp of the report (to reflect the latest time at which the cached object is known to be valid). Note that Broadcast Timestamp employs cache-level invalidation, i.e., regardless of whether there is a pending query to the object or not, every cached object is checked against the report. Figure 6.3 shows the algorithmic description of the scheme.

EXAMPLE 9 *Consider the running example (see Example 8). Suppose the server broadcasts the OIR at an interval of 2 time units (i.e., L = 2), and maintains updates for the last 6 time units (i.e., $w = 3$). Suppose a client has the following cache content:* $[(o_4, 28), (o_6, 28), (o_{12}, 28)]$ *when it disconnects at time 29. Note that the last report received is at time 28. Let the current time be $T = 33$ when the client reconnects. It receives the following report at time 34:* $\{34, [(o_8, 32), (o_{12}, 30), (o_{16}, 28)]\}$. *From the report, the client can*

```
// T – timestamp of current report
// T^c – timestamp of last valid report received by mobile client
// t^c_id – timestamp of cached object O
if (T − T^c > wL)
    discard the entire cache
else {
    for every object O in the client cache {
        if there is a pair [o_id,t_id] ∈ OIR {
            if (t^c_id < t_id)
                invalidate O and remove it from the cache
            else
                t^c_id ← t_id
        }
    }
    for every object O ∈ Q {
        if O is in the cache
            use the cache's content to answer the query
        else
            submit request for O
    }
    T^c ← T
}
```

Figure 6.3. The Broadcast Timestamp scheme.

invalidate object o_{12} while updating the timestamp of the valid objects, o_4 and o_6, to 34. □

6.2.2 GCORE

The Broadcast Timestamp's effectiveness in salvaging the cache content is largely dependent the duration of the disconnection. When the client disconnects longer than the duration of the log maintained at the server, i.e., client disconnection time is greater than wL time units, the entirety of the cache will be discarded. The GCORE (Grouping with COld update-set REtention) scheme [123] was proposed to avoid discarding the entirety of the cache content after a long disconnection.

GCORE is a group-based strategy and adopts the following framework:

Content. The report consists of a list of (id,TS) pairs. The size of the report varies according to the fixed update history. There is no selective tuning.

Invalidation Mechanism. The invalidation is performed by both the client and the server. The client's cache invalidation is done at the cache-level.

Log. The server maintains logs for objects. The size of the log varies though the log history is fixed at $[T - WL,\ T]$ for objects, where T is the current time, and W is the update log window, and L is the fixed interval at which the reports are broadcast. We note that the server also keeps, for each group of objects, the timestamp of the most recent update to the group.

In GCORE, the server organizes the whole database into a set of disjoint groups, each containing the same number of objects. Note that each data object has an additional attribute (i.e., the group id) associated with it.[2]

Let the current time be T, and the timestamp of the last report received by a client be T^c. Like the Broadcast Timestamp scheme, the server broadcasts an object invalidation report that contains the update history of the past w broadcast intervals, and is obtained from the information maintained between $T - wL$ and T. The contents of OIR are the current timestamp T and a list of (o_{id}, t_{id}) pairs where o_{id} is an object identifier of an object updated during the interval $[T - wL,\ T]$ and t_{id} is the corresponding most recent update timestamp and $t_{id} > (T - wL)$.

The server maintains for each group the following information: a list of object ids and their most recent update timestamps for objects updated in the interval $[T - WL,\ T]$ $(W > w)$, the most recent update time of the group, and a counter that keeps the number of distinct objects that were updated during the interval $[T - WL,\ T - wL]$. The counter is set as such because of the following observation to salvage some valid objects: if no object was most recently updated in the interval $[T - WL, T - wL]$ and $T^c > (T - WL)$, it means that if there are any objects in the group updated since T^c, their ids would have been included in the invalidation report; thus, those objects that are not invalidated by the invalidation report are valid and can be salvaged.

When a mobile client reconnects and queries the database, it first tunes to the downlink channel and listens for the next object invalidation report. From the OIR, the client can invalidate its cached objects that are found in the OIR. These objects are then *removed* from the cache. For the remaining objects that are in the cache, the client will submit their group ids to the server for a group validity check by the server.

The server checks the validity of a group by examining whether or not all the objects updated since the client becomes disconnected have been included in the latest invalidation report. If yes, then the group can be retained by the client. Otherwise, the entire group is invalid. This can be achieved by first checking if the group timestamp is less than T^c. If yes, the group is valid since the group is last updated before the mobile client becomes disconnected. If not, the server checks if the counter equals to 0 and $T^c > (T - WL)$. If yes, this group is also valid since the updated object ids are all included in the invalidation report. Otherwise, this group is invalid. The client is then informed of the validity of the groups.

The combined client and server protocols for the GCORE scheme is shown in Figure 6.4. We note that if the client disconnection time is longer than WL, then the entirety of the cache content has to be discarded.

EXAMPLE 10 *We shall use the running example to illustrate GCORE. Let* $w = 3$, $W = 6$ *and* $L = 2$. *Suppose a mobile client disconnects at time* $t_d = 23$ *(where* $T > t > T - WL$*) and reconnects at time* $t_r = 33$. *The next invalidation report that the client receives is at time* $T = 34$. *Thus, we have* $T > t_r > T - wL > t_d > WL$. *Suppose the client generates a query that requests objects* $o_2 - -o_8$, o_{12} *and* o_{16} *that are in its cache and belong to the following three groups:* $G_1 = \{o_2, o_3, o_{12}, o_{16}\}$, $G_2 = \{o_4, o_5, o_8\}$ *and* $G_3 = \{o_6, o_7\}$. *At time* T, *the OIR's content is* $\{34, [(o_8, 32), (o_{12}, 30), (o_{16}, 28)]\}$. *At the server, the values of the counter for* G_1, G_2 *and* G_3 *are 0, 0 and 1 respectively. From the OIR, the client invalidates* o_{12}, o_{16} *from* G_1, *and* o_8 *from* G_2. *Thus, these objects can be "removed" from the cache, and the remaining objects whose validity are unknown are given by the resultant groups:* $G_1 = \{o_2, o_3\}$, $G_2 = \{o_4, o_5\}$ *and* $G_3 = \{o_6, o_7\}$. *The client submits the group ids of* G_1, G_2 *and* G_3 *to the server for group validity check. The server found that the counter values for* G_1 *and* G_2 *are both 0, and so objects of these groups in the cache will be valid.* G_3, *having a counter value of 1, is found to be invalid, and so all objects of* G_3 *in the cache (i.e.,* o_6 *and* o_7*) will be discarded.* □

6.2.3 Dual-Report Cache Invalidation

Like GCORE [123], the Dual-Report Cache Invalidation scheme (DRCI) [26][3] employs the group-oriented logging approach and can reduce unnecessary downlink costs due to group cache misses. However, it avoids the excessive uplink requests (and hence high energy consumption) required by GCORE especially if the data cached in the mobile clients belong to many different groups. This is achieved by employing an exclusively broadcast-based approach to minimize uplink requests, so that the server does not need to participate in the invalidation process. It is based on the following design spaces:

Content. The report consists of a list of (id,TS) pairs and a list of (group-id,TS) pairs. The size of the report varies with a fixed update history at wL time units. There is no selective tuning.

Invalidation Mechanism. The invalidation is at the cache-level, and is performed by the client only.

// T – timestamp of current report
// T^c – timestamp of last valid report received by mobile client
// t^c_{id} – timestamp of cached object O

Client protocol:

if $(T - T^c > WL)$
    discard the entire cache
else {
    for every object O in the client cache {
        if there is a pair $[o_{id}, t_{id}] \in$ OIR {
            if $(t^c_{id} < t_{id})$
                invalidate O and remove it from the cache
            else
                $t^c_{id} \leftarrow t_{id}$
        }
    }
    if $(T - T^c > wL)$ {
        for every group G_{gid} in the client cache that is non-empty {
            submit T^c aad G_{gid} to server
            invalidate the entirety of group G_{id} if the server says so
        }
    }
}
for every object $O \in Q$ but not in the cache
    submit request for O
answer the query with the cached objects
$T^c \leftarrow T$

// T_{id} - Group G_{id}'s last update timestamp at server
// c_{id} - counter associated with G_{id} at server

Server protocol:

for each group G_{id} queried by client {
    if $(T_{id} < T^c)$
        G_{id} is valid
    else {
        if $(c_{id} == 0 \wedge T^c > (T - WL))$
            G_{id} is valid
        else
            G_{id} is invalid
    }
}

Figure 6.4. The GCORE scheme.

Log. The server maintains logs for objects. The size of the log varies though the log history is fixed at [T-WL,T] for objects, where W is the update log window.

Under DRCI, the server broadcasts every L time units a pair of invalidation reports, an *object invalidation report* (OIR) and a *group invalidation report* (GIR). Let the current timestamp be T. To generate the reports, the server keeps track of all the objects updated between $T - wL$ and T, and between $T - WL$ and T, where w and W are the broadcast and update log windows respectively, and $W > w > 0$. Note that the distinction on the information maintained between the two time intervals is only a logical one. Clearly, objects appearing in the interval $[T - wL,\ T]$ will also appear in the interval $[T - WL,\ T]$, and in an actual implementation, such redundancies can be avoided altogether. The most recent OIR broadcast contains the update history of the past w broadcast intervals and is obtained from the information maintained in the interval $[T - wL,\ T]$. The contents of OIR are the current timestamp T and a list of (o_{id}, t_{id}) pairs where o_{id} is an object identifier of an object updated during the interval $[T - wL,\ T]$ and t_{id} is the corresponding most recent update timestamp and $t_{id} > (T - wL)$. The GIR, on the other hand, is a fixed-size report that contains the update history of the past W broadcast intervals and is obtained from the information maintained in the interval $[T - WL,\ T]$. GIR contains for each group a timestamp that reflects the most recent timestamp in which the group is valid. In other words, the contents of the GIR is a list of (G_{id}, T_{id}) pairs where G_{id} is a group identifier and T_{id} is the most recent timestamp in which the group G_{id} is valid.

The timestamps of the groups in GIR are determined only at the time when the invalidation reports are to be broadcast and are assigned as follows:

- **Step 1.** Based on the observation that objects that are updated during $[T - wL,\ T]$ would be reflected in the OIR, these objects are excluded from consideration when determining the timestamp of a group.
- **Step 2.** For each group, perform the following. Among the group's remaining objects in the update history, find the one with the largest timestamp that is less than $T - wL$. Let this timestamp be t. If no such objects exists (i.e., no updates during $[T - WL,\ T]$ or the updates are already included in the OIR), then $t = 0$. The timestamp for the group is $\max(T - WL,\ t)$.

The DRCI scheme facilitates invalidation in two ways. First, for those clients with a short disconnection time, the OIR can be used solely to invalidate the cache content. Second, for those clients disconnected before time $T - wL$, both the OIR and GIR are used together — the OIR is used to invalidate individual objects in the cache first, the GIR is then used to invalidate the remaining objects whose groups have been updated. Since the GIR exclude objects in

OIR, the OIR effectively helps to minimize false invalidation. The GIR can further minimize invalidation of the entire cache for long disconnection time. The algorithmic description of the protocol is shown in Figure 6.5. Note that the entire cache content will be discarded if the client disconnection time is longer than $T - WL$ time units.

```
// T – timestamp of current report
// T^c – timestamp of last valid report received by mobile client
for every pair [o_id, t_id] ∈ OIR {
    if (object o_id is in the client cache) {
        if (T^c < t_id )
            remove object from the cache
    }
}
if (T – T^c > wL) {
    for every group G_id in the client cache {
        examine the corresponding pair [G_id, T_id] in GIR
        if (T^c < T_id)
            remove all objects in group from the cache
    }
}
for every object O ∈ Q {
    if (O is in the cache)
        use the cache's content to answer the query
    else
        submit request for O
}
T^c ← T
```

Figure 6.5. The Dual-Report Cache Invalidation scheme.

Although the basic DRCI scheme is independent of the way the database objects are grouped, its performance can be influenced by the grouping scheme. DRCI distinguishes four categories of objects: hot update-hot demand (HH), hot update-cold demand (HC), cold update-hot demand (CH) and cold update-cold demand (CC). Clearly, it makes no sense to group objects in the HH category (hot update-hot demand) with objects in the CC category (cold update-cold demand). Therefore, groups are formed from objects in the same category.

There is also a tradeoff in the number of groups. If the number of groups is too large, the size of the GIR will also be large, and this will utilize a significant amount of the channel bandwidth which may lead to poor channel utilization. On the other hand, a small number of groups may also be bad. This is because when the number of groups is small, the size of the groups will be large. As a result, it becomes more likely for groups to be invalidated.

EXAMPLE 11 *We shall use the running example to illustrate DRCI. Let* $w = 3$, $W = 6$ *and* $L = 2$. *Suppose a mobile client, MC, disconnects at time* $t_d = 23$ *(where* $T > t > T - WL$*) and reconnects at time* $t_r = 33$. *The next invalidation report that the client receives is at time* $T = 34$. *Thus, we have* $T > t_r > T - wL > t_d > WL$. *Suppose MC generates a query that requests objects* $o_2 - -o_8$, o_{12} *and* o_{16} *that are in its cache and belong to the following three groups:* $G_1 = \{o_2, o_3, o_{12}, o_{16}\}$, $G_2 = \{o_4, o_5, o_8\}$ *and* $G_3 = \{o_6, o_7\}$. *At time T, we have the following: the contents of OIR are* $\{34, [(o_8, 32), (o_{12}, 30), (o_{16}, 28)]\}$, *and the contents of GIR for groups* G_1, G_2 *and* G_3 *at time T are* $\{(G_1,\ 22), (G_2,\ 22), (G_3,\ 26) \cdots\}$. *From the OIR, MC invalidates* o_{12}, o_{16} *from* G_1, *and* o_8 *from* G_2. *Thus, these objects can be "removed" from the cache, and the remaining objects whose validity are unknown are given by the resultant groups:* $G_1 = \{o_2, o_3\}$, $G_2 = \{o_4, o_5\}$ *and* $G_3 = \{o_6, o_7\}$. *Since* $T - WL < t_d$ *and* $t_d < 26$, *therefore the remaining objects in groups* G_1 *and* G_2 *are considered valid, while all the objects in* G_3 *are considered invalid.* □

6.2.4 Bit-Sequences

The Bit-Sequences (BS) algorithm [66] adopts the following framework:

Content. The report consists of a list of (list of ids, TS) pairs in a compact form. This allows the report size to be fixed, though the update history varies. There is no organization to support selective tuning.

Invalidation Mechanism. The invalidation is performed by the client at cache-level.

Log. The logs maintained at the server keeps track of individual object update information (using (id,TS) pairs) for up to half the database size, i.e., the size of the log is fixed but the update history is variable.

Let the number of database objects be $N = 2^n$. In the BS algorithm, the invalidation report reflects updates for n different times — T_n, T_{n-1}, ..., T_1, where $T_i < T_{i-1}$ for $1 < i \leq n$. The report comprises n binary bit-sequences, each of which is associated with a timestamp. Each bit represents a data object in the database. A "1" bit in a sequence means that the data object represented by the bit has been updated since the time specified by the timestamp of the sequence. A "0" bit means that the object has not been updated since that time. The n bit-sequences are organized as a hierarchical structure with the highest level (i.e., bit-sequence B_n) having as many bits as the number of objects in the database, and the lowest level (i.e., bit-sequence B_1) having only 2 bits. For the sequence B_n, as many as half of the N bits (i.e., $N/2$) can be set to "1" to indicate the $N/2$ objects that have been updated (initially, the number of "1" bits may be less than $N/2$). The timestamp of the sequence B_n is T_n. The next sequence in the structure, B_{n-1}, has $N/2$ bits. The kth bit in B_{n-1} corresponds to the kth "1" bit in B_n, and $N/4$ bits can be set to "1" to indicate that $N/4$ objects have been updated since T_{n-1}. In general, for sequence B_{n-i}, $0 \leq i \leq n-1$, there are $N/2^i$ bits, and the sequence will reflect that $N/2^{i+1}$ objects have been updated after the timestamp T_{n-i}. The kth bit in sequence B_{n-i} corresponds to the kth "1" bit in the preceding sequence (i.e., B_{n-i+1}). An additional dummy sequence B_0 with timestamp T_0 is used to indicate that no object has been updated after T_0.

In general, N does not need to be a power of 2, and the number of lists can also be any value other than n. Furthermore, the list associated with timestamp T_i does not need to reflect the updates for half the number of objects in the list associated with T_{i+1}, $1 \leq i \leq n-1$. However, for simplicity, we have chosen to follow the approach proposed in [66].

The Bit-Sequences structure is broadcast to clients periodically. The protocol for invalidating the cache is shown in Figure 6.6.

EXAMPLE 12 *Figure 6.7 shows the invalidation report represented by the Bit-Sequences structure for our running example. As shown, the first level (B_4) has 16 bits, 8 of which has been set to "1". These 8 objects are the most recently updated objects. The timestamp for B_4 is 18. Similarly, bit sequence B_1 has 2 bits, and reflects the most recently updated object 8, and has timestamp of 32. Assume that the client receives this invalidation report when it submits its query for objects 5 and 8. Suppose the last invalidation report received by the client before it disconnects is at time 31. Since the client's cache content is last valid at time 31, it should use the sequence B_2 to invalidate its cache. To locate those objects denoted by the two "1" bits, the client will check the sequences B_2—B_4. This is accomplished as follows. To locate the object corresponding to the second bit that is set to "1" in B_2, the client has to check the second "1" bit in B_3. Since the second "1" bit in B_3 is in the 5th position, the client will have to examine the 5th "1" bit in B_4. Because B_4 is the highest bit-sequence, and the 5th "1" bit is in the 8th position, the client can conclude that object 8 has been updated since time 31. Similarly, the client can determine that the 12th object has also been updated since that time. Therefore, both objects will be invalidated by the client. Since the client requests for objects 5 and 8, object 5 remains valid and can be used to answer the query, while the request for the invalid object 8 has to be submitted to the server.* □

```
// T – timestamp of current report
// T^c – timestamp of last valid report received by mobile client
if T_0 ≤ T^c
    all cached objects are valid
else {
    if T^c < T_n
        remove the entire cache content
    else {
        determine the bit sequence B_i such that
            T_i ≤ T^c < T_{i-1}, 1 ≤ i ≤ n
        invalidate all the objects marked "1" in B_i
    }
}
for every object O ∈ Q {
    if (O is in the cache)
        use the cache's content to answer the query
    else
        submit request for O
}
T^c ← T
```

Figure 6.6. The Bit-Sequences scheme.

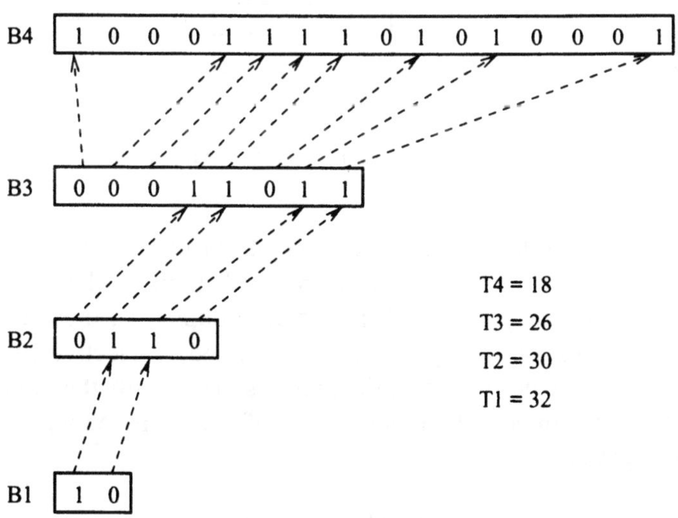

Figure 6.7. The Bit-Sequences structure.

6.2.5 Selective Dual-Report Cache Invalidation

Scheme DRCI is, unfortunately, not energy efficient as there is no mechanism to support selective tuning of the invalidation report, and the invalidation is performed by client at cache-level. In other words, the entirety of the report has to be downloaded by the clients, and every object in the report examined.

An energy-efficient version of DRCI, the Selective Dual-Report Cache Invalidation scheme (SDCI), was proposed in [26][4]. SDCI differs from DRCI in two ways: the report is organized to facilitate selective tuning. and the invalidation is query-level-based. SDCI organizes the pair of invalidation reports (the OIR and GIR) as follows. First, the GIR is broadcast before the OIR. Second, the entries in the OIR are ordered and broadcast based on the groups, i.e., updates in the same group will appear together. A partition symbol will separate continuous groups. Third, an additional "pointer" is added to each element of the GIR, and this pointer reflects the starting position of the objects within this group in the OIR. Thus, the GIR consists of triplets of the form (group-id, TS, ptr) where group-id represents the group identifier, TS is the timestamp of the most recent update (excluding those in the OIR) of the group, and *ptr* is an offset to the starting position of the objects in OIR corresponding to the group identified by group-id. Figure 6.8 shows the organization of the invalidation reports.

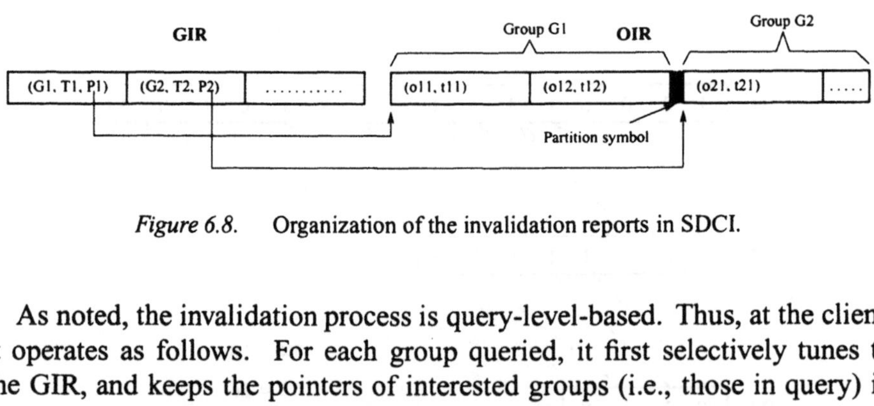

Figure 6.8. Organization of the invalidation reports in SDCI.

As noted, the invalidation process is query-level-based. Thus, at the client, it operates as follows. For each group queried, it first selectively tunes to the GIR, and keeps the pointers of interested groups (i.e., those in query) in memory. Once all the desired groups are determined, it selectively tunes to the respective position in the OIR using the pointers. Thus, only the desired groups and objects are examined. The functionality of the improved algorithm SDCI is shown in Figure 6.9.

6.2.6 Bit-Sequences with Bit Count

The Bit-Sequences with Bit Count (BB) method is an energy-efficient scheme based on the Bit-Sequences algorithm [106]. Like the Bit-Sequences approach, it comprises a set of bit sequences organized in a hierarchical man-

```
// T^c_id – last known timestamp of group with id G_id
// t^c_id – last known valid time of object with id o_id
P ← φ
for (every group queried) {
        selectively tune to the triplet [G_id, T_id, P_id] in GIR_i
        if T^c_id < T_id
            invalidate all objects of group
        else {
            T^c_id ← T_id
            P ← P ∪ {P_id}
        }
}
while (P ≠ φ) {
        p ← first(P)
        P ← P − {p}
        switch to doze mode until the position p is coming
        while (not encounter partition symbol) {
            download pair [o_id, t_id]
            if (o_id is in client cache) {
                if t^c_id < t_id
                    invalidate object
                else
                    t^c_id ← t_id
            }
        }
}
for (every object O ∈ Q ) {
        if (O is in the cache)
            use the cache's content to answer the query
        else
            submit uplink request for O
}
```

Figure 6.9. The Selective Dual-Report Cache Invalidation scheme.

ner. However, only the relevant bits need to be examined. This is achieved by associating each bit sequence with a bit count array.

Let $N = 2^n$ be the number of objects in the database. Suppose that a query, Q, returns the set of objects $\{O_1, O_2, \ldots, O_q\}$ as answers, and that these objects are already ordered in the same manner as the information reflected in the invalidation report, i.e., information for O_1 will be received before information for O_2, and so on. The corresponding timestamps at which these objects are last valid (in the client cache) are denoted as $t_1, t_2, \ldots, t_q$ respectively.

As in the BS scheme, the BB structure comprises a set of n bit sequences: sequence B_n has a timestamp T_n that indicates that updates after T_n are reflected, and comprises N bits, half of which are set to "1"; sequence B_{n-1} has timestamp T_{n-1} and $N/2$ bits, of which $N/4$ bits are set to "1"; and so on. In fact, the content of the bit sequences are exactly the same as those of the BS scheme. Like the BS scheme, if the bit sequence B_{n-i} is to be used to invalidate the cache, then the sequences $B_{n-i}, B_{n-i+1}, \ldots, B_n$ *may* have to be examined. However, strategy BB adopts a *top-down* examination of the sequences, i.e., from B_n to B_{n-i}, rather than the *bottom-up* approach (i.e., B_{n-i} to B_n) of BS scheme. Moreover, for some valid objects, it may not be necessary to examine all the sequences from B_n to B_{n-i} as it may be possible to determine their validity and terminate the search before sequence B_{n-i}. Furthermore, the scheme only examines the relevant bits in each sequence. As the kth "1" bit in B_{n-i} corresponds to the kth bit in B_{n-i-1}, a mechanism that can count the number of "1" bits in a sequence without examining the entire sequence is needed. Before we discuss this mechanism, let us illustrate how selective tuning can be facilitated with such a mechanism. To validate an object O, the client first identifies the bit sequence that should be used. This is accomplished by examining the set of timestamps. Suppose this is sequence B_{n-i}. This means that the sequences B_n has to be examined, followed by B_{n-1}, and so on till B_{n-i}. From object O, the client can selectively tune to the corresponding bit in B_n (without scanning the entire B_n). If the bit is set to "0", then the object is valid (since object O will not be found in any subsequent sequences $B_{n-1}, B_{n-2}, \ldots$); otherwise, the client determines (to be discussed shortly how this can be done) the number of "1" bits from the beginning of B_n to the bit corresponding to O. From this number, it can again selectively tune to B_{n-1}, and examine the corresponding bit of O in B_{n-1}. Again, if the bit is "0", then the object O is valid and the search terminates; otherwise, its position in B_{n-2} is determined and this process is repeated until sequence B_{n-i}. The process terminates when the "0" bit is encountered at any of the sequences from B_n to B_{n-i}. If the relevant bit at B_{n-i} is "1", then the object is invalid; otherwise, it is valid.

Now, the mechanism to facilitate selective tuning is simple. Basically, each bit sequence has associated with it a bit count array, all of which have entries that are j bits. For bit sequence B_{n-i}, $0 \le i \le n-1$, the sequence is partitioned into packets of 2^j bits. In other words, there are $\lceil (N/2^i)/2^j \rceil$ packets. In general, for sequence B_{n-i}, the number of array entries is $\lceil (N/2^i)/2^j \rceil$. We note that sequences with fewer than j bits do not need to be associated with a bit count array as all bits have to be examined anyway. Essentially, the kth entry in the bit count array of sequence B_{n-i} represents the number of "1" bits that have been set for the kth packet in the sequence. Selective tuning is achieved as follows. The client knows which bit/position in a sequence it is looking for. Let packet i contains the bit that the client is interested in. From the bit array count, the client can determine the number of "1" bits that have been set for packets 1 to $i - 1$. The client can then tune into the ith packet, and scan the ith packet until the relevant bit. In this way, the number of "1" bits can be determined.

To check the validity of the answer objects to query Q, the client employs the protocol shown in Figure 6.10.

The invalidation report is organized as follows. The counter is broadcast first, the timestamps are broadcast next, followed by the bit count arrays for sequences B_n, B_{n-1}, ... and finally the bit sequences B_n, B_{n-1}, ..., B_1.

EXAMPLE 13 *We illustrate scheme BB with the same running example in Figure 6.2. The organization of the invalidation report for BB is shown in Figure 6.11. In this example, each bit count array entry keeps track of the number of "1" bits set for 4 objects. Since there are 16 objects in the database, there are 4 entries in the bit count array corresponding to B_4, 2 entries in B_3's bit count array, and 1 entry in B_2's bit count array. We note that B_1 is not associated with a bit count array. As before, assume that a query requests for objects 5 and 8, whose cached timestamps are respectively 31 and 27. From the timestamps in the invalidation report, the client knows that it needs to check B_2 for the validity of object 5, and B_3 for the validity of object 8. The client first determines which two bits in B_3 correspond to the two queried objects. This is done as follows. As both objects 5 and 8 are in the same packet, from the first bit count array entry (of B_4), the client knows that there is only 1 "1" bit among the first 4 objects in the bit-sequence B_4. Thus, it will tune to the beginning of the second packet of B_4, and examine the first bit (in the second packet) till the fourth bit. Since the first bit corresponds to the object 5 and it is set to "1", the client knows that object 5 is the second bit in B_3. Similarly, the client can determine that object 8 is the fifth bit in B_3. For object 5, the client examines the corresponding bit in B_3 which has been set to "0" indicating that the object is valid. For object 8, the client first examines the bit count array for B_3, and knows that the first entry contains a value of 1. By examining the first bit of the second packet of B_3, it determines that the bit corresponding to object 8 is set to "1". This also means that object 8 can be found in the second bit of B_2. It then examines the second bit of B_2 and note that object 8 is invalid.* □

6.2.7 Discussion

Earlier work has shown that BGI, a variation of DRCI, is more energy efficient than GCORE without sacrificing on the access time performance [107]. BGI differs from DRCI in the way objects are grouped. BGI considers

```
// AnswerSet = {o_1, o_2, ..., o_q} — objects for query Q
// t_id – last known valid time of object with id o_id
download the counter, the timestamps and the bit count arrays
for each object o_i ∈ AnswerSet
    if T_O ≤ t_i // object o_i is valid
        AnswerSet = AnswerSet - {o_i}
    else if t_i < T_n // object o_i is invalid
        AnswerSet = AnswerSet - {o_i}
// AnswerSet contains the remaining objects whose validity is still uncertain
for each object o_i ∈ AnswerSet
    determine the bit sequence to be used to validate o_i
k = n
repeat {
    for each object o_i ∈ AnswerSet, examine bit sequence B_k {
        tune to packet p containing information on o_i
        examine the bits in packet p until position of o_i
        let the number of "1" bits set (in p inclusive of o_i) be b
        if B_k is the bit sequence to be used to validate o_i {
            if the bit corresponding to o_i is set to "1"
                o_i is invalid
            else
                o_i is valid
            AnswerSet = AnswerSet - {o_i}
        } else {
            if the bit corresponding to o_i is set to "0" {
                o_i is valid
                AnswerSet = AnswerSet - {o_i}
            } else {
                // we need to examine the next sequence B_{k-1}
                from the bit count array of B_k
                    determine the number of "1" bit from
                        packets 1 to p − 1 of B_k
                let this value be c
                the position of o_i in bit sequence B_{k-1} is (c + b)
            }
        }
    }
    k = k − 1
} until AnswerSet = ∅
```

Figure 6.10. The Bit-Sequences with Bit Count scheme.

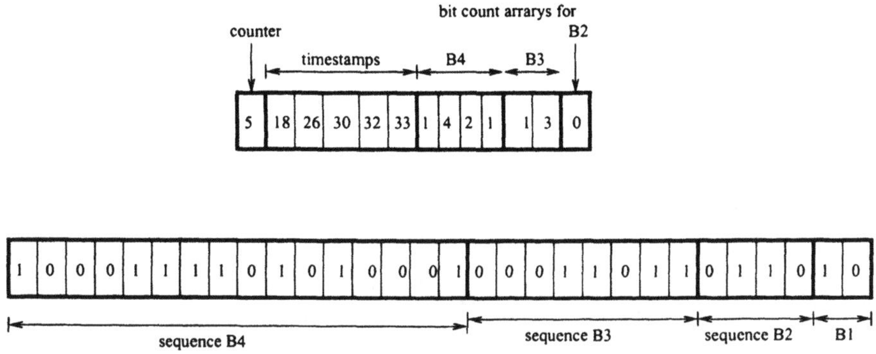

Figure 6.11. The Bit-Sequences with Bit Count method.

only two categories (hot-update and cold-update) while DRCI divides the database into four categories. The performance study in [26] also reported that the Broadcast Timestamp scheme performs poorly in terms of both access time and tuning time when compared against the DRCI and SDCI schemes. When the disconnection time is long, under Broadcast Timestamp, the entirety of the client's cache content has to be discarded even if they are actually valid. On the other hand, DRCI and SDCI can both salvage the cache content as much as possible.

In [108], a simulation study was conducted to evaluate four schemes: Dual-Report Cache Invalidation (DRCI) and Bit-Sequences (BS), and their energy-efficient counterparts, Selective Dual-Report Cache Invalidation (SDCI) and Bit-Sequences with Bit-Count (BB). The scheme where the client queries the server for the validity of the cache is also used as a reference. This scheme is denoted SI. We shall report here two interesting sets of results. The schemes are compared on two metrics: access time and energy consumption. There, energy consumption is measured in terms of *energy units*. Each unit is defined as the energy consumed on receiving 1 Kbits of information. Moreover, it was assumed that transmitting data is 10 times more power consuming than receiving data [45]. The study only measures the energy expended in receiving the report, submitting the queried objects and receiving the answer objects.

Effect of Query Arrival Time

In this study, the number of concurrent users that can be supported (i.e., the scalability) was investigated. This is important as the larger the number of queries the environment can support, the better the scheme is. The query arrival time was varied from 0.1 to 20 sec. The results of this study is shown in Figure 6.12.

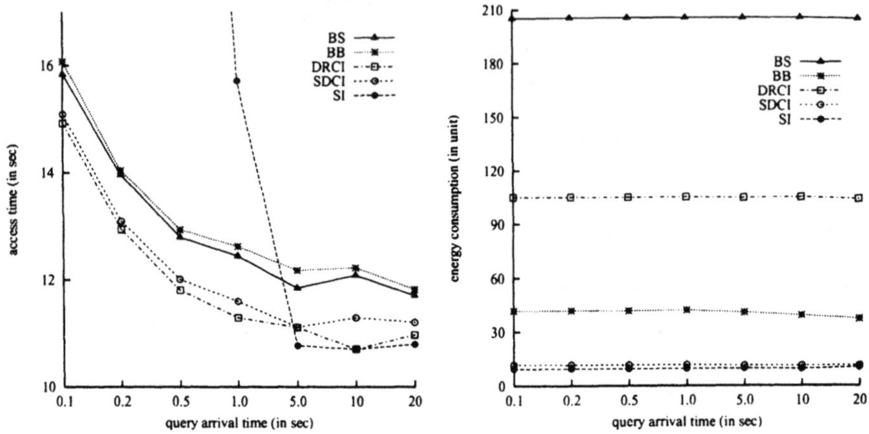

(a) Access time

(b) Energy consumption

Figure 6.12. Effect of query arrival time.

We shall first look at the results of the four schemes: BB, BS, DRCI and SDCI. For the four schemes, the access time decreases with increasing query arrival time. This follows from the fact that with larger arrival time, the number of concurrent queries in the system is smaller and the bandwidth is sufficient to handle these queries. However, when the query arrival time is small, the downlink channel becomes congested as more queries are issued. The contention for the limited bandwidth leads to the higher access time. We also note that the energy-efficient schemes are slightly worse than their counterparts that do not facilitate selective tuning, i.e., BB is worse than BS and SDCI is worse than DRCI. As BB and BS reflects the same update information in the invalidation reports, this result is expected by virtue of the fact that the report size of the BB scheme is larger. The same reason applies to schemes SDCI and DRCI. We also observe that the Bit-Sequences methods are worse than the Dual-Report approaches. This can be explained as follows. First, the report size of the Bit-Sequences methods is larger than that of the Dual-Report approaches, thus consuming more bandwidth (for this experiment, the report size of the Bit-Sequences methods is more than twice as large). Second, when the disconnection time is small, the OIR of the Dual-Report methods can be used to invalidate stale objects and false invalidations are minimal; when the disconnection time is high, all methods have about the same amount of invalidations.

In terms of energy consumption, we note that all the schemes are largely unaffected by query arrival time. This is so since the energy consumption comes from examining the invalidation report, submitting the queries and receiving the results of invalidated objects, and all these components are independent of the query arrival time. As expected, the energy efficient schemes outperform their counterparts by a wide margin (more than 10 times more energy efficient). Since the BB scheme has to examine the bit count arrays, it consumes more energy than the SDCI scheme.

From the access time result, we note that scheme SI performs best when the inter-arrival time is large. This is because the system load is light. More importantly, all the other four approaches have a smaller effective downlink bandwidth (part of the bandwidth is used to broadcast the invalidation reports) compared to scheme SI. However, when the inter-arrival time becomes small, scheme SI's need to submit the requested objects' information results in the server and the uplink channel becoming a bottleneck very quickly. The other approaches, on the other hand, only submit invalid objects' information. In terms of energy consumption, we note that scheme SI's energy consumption is fairly good. In fact, it is the best among all the schemes. While it incurs more energy in transmitting objects' information, it does not need to examine invalidation reports. The net effect in this experiment is an overall better performance. However, because of its poor access time performance, scheme SI is expected not to scale well, and has not received much attention.

Effect of Update Arrival Time

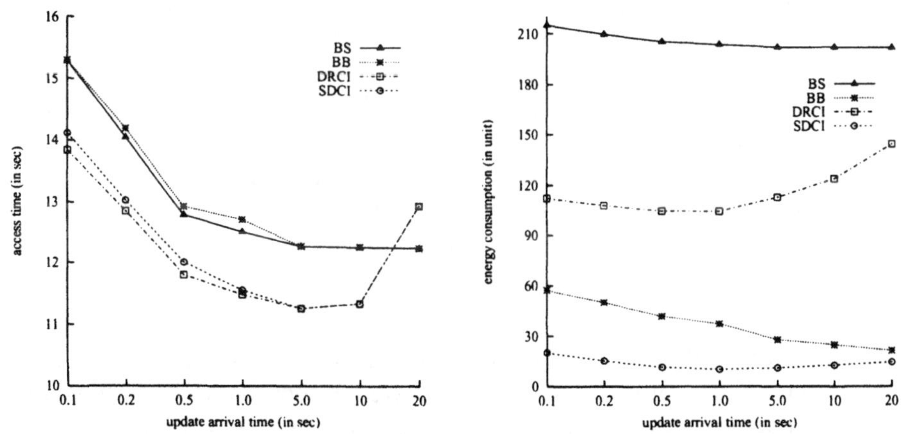

(a) Access time

(b) Energy consumption

Figure 6.13. Effect of update arrival time.

In this study, the effect of update arrival time was examined by varying the update arrival time from 0.1 to 20 sec. Figure 6.13 shows the effect of update time on access time and energy consumption. First, we observe that the relative performance between BS and BB, and DRCI and SDCI, is similar to those results presented earlier, i.e., BS and DRCI are respectively slightly better than BB and SDCI in terms of access time, but are respectively less energy efficient than their counterparts in terms of energy consumption.

Second, for the Bit-Sequences-based approaches, we note that as the update time increases, the access time and energy consumption also reduce. A shorter update time means more objects being updated in a shorter timespan. As a result, the difference between timestamps in the invalidation report is small. This implies that updated objects associated with a smaller timestamp value (those further away from the current time) are the ones that the client is likely to be looking out for. Recall that in the invalidation report, a smaller timestamp would be associated with a larger number of updated objects. Thus, more objects are being invalidated, and the client needs to refresh more objects. This explains why both the access time and energy consumption are higher. On the other hand, a longer update time means that the difference between timestamps in the report is large, and so fewer objects are likely to be invalidated. Thus, the access time and energy consumption are smaller compared to that of shorter update time.

Third, for the Dual-Report-based approaches, we see a slightly different picture. There exists a certain update rate at which the system performs best, i.e., too low or too high an update rate can degenerate the system performance. A high update rate implies a high probability of cache invalidation, and also a larger update history for a fixed w. This results in a large OIR that consumes the downlink bandwidth and hence the poor performance. On the other hand, when the update rate is low, the probability of false invalidation increases, as the updated object may no longer appear in the OIR but reflected in the GIR.

Finally, in terms of energy consumption, we note that the relative performance between the four schemes remains largely the same. For the Dual-Report schemes, we note that there is also an "optimal" point. This follows from the same reasons as the case of the access time performance.

Notes

1 The original algorithm in [15] maintains and broadcasts updates in the last ω time units. While ω and L are unrelated except that $\omega \geq L$, most of the subsequent works assume that $\omega = wL$. For consistency, we have also done so for all algorithms presented in this chapter.

2 Alternatively, hashing techniques can be employed, in which case, objects hashed to the same partition belong to the same group. Here, no additional attribute is necessary but each client needs to know the hashing functions. Furthermore, groups may not contain the same number of objects.

3 In [26], the scheme is referred to as Hybrid Cache Invalidation.

4 In [26], the scheme is referred to as Selective Cache Invalidation.

Chapter 7

BALANCING PUSH AND PULL

Traditionally, there are two basic modes to disseminate information in a *dissemination-based information system* (DBIS) [46]. In the first mode, which is the *push* (or data publishing) mode, data objects are periodically broadcast by the server based on some *broadcast program*, and clients simply need to tune to the communication channel, listen to the data transmitted, and filter the stream for the desired data objects. In the second mode, known as the *pull* (or demand-driven) mode, clients explicitly request for data objects through the communication channel and the server responds by sending back the requested objects. While the pull mode has been the primary mode of information dissemination in client-server architecture, the push mode has found some amount of acceptance in *asymmetric communication environments*.

As early as the late 80s, it has been shown that neither of the two modes are superior [121]. The pull model is effective when the number of clients is small. However, it does not scale well: as the number of clients increases, the channel bandwidth and the server become a bottleneck quickly. Even if the infrastructure were developed to meet the demand in both network bandwidth and server capacity, the approach is still uneconomical as most of this infrastructure would be underutilized and wasted during non-peak periods. On the other hand, data disseminated under the push model is independent of the number of clients listening to the broadcast. For effective performance, this method requires that the access patterns be identified to determine an optimal broadcast program. But, it is unable to capture the changing access patterns of users — it has no mechanism to detect "hits" or "misses". Therefore, integrating the two modes for information dissemination is expected to reap much benefits.

In this chapter, we shall present integrated schemes that effectively combine both the push and pull models to disseminate information. Section 7.1 presents the architecture of an integrated model for information dissemination. In

Section 7.2, we examine the case for *dynamic* hybrid delivery mechanisms. Finally, in Section 7.3, we shall present several promising dynamic schemes that have been proposed in the literature.

7.1 ARCHITECTURE OF INTEGRATED MODEL

Figure 7.1 shows the generic architecture of an integrated model that disseminates data using both the push and pull modes. The server generates a broadcast program for frequently accessed objects. These objects are periodically broadcast according to the program. The server also services on-demand requests from clients. At any point in time, the server either broadcasts an object from the broadcast program or as a response to a queued object request. For ease of presentation, we shall refer to the downlink channel allocated for the pushed data as the *push channel* and that allocated for the pulled data as the *pull channel*. Clients can either listen to the broadcast to filter their desired objects without submitting any requests, or submit requests on the uplink channel and wait for the requests to be serviced and the object disseminated. Designing an integrated model calls for the following issues to be addressed:

- How can we determine "hot" and "cold" data? Which are the objects to be periodically broadcast, and which should be disseminated on demand?
- How should the pushed and pulled data be interleaved? In other words, how can we optimize the allocation of bandwidth to the pushed and pulled channels?
- How can we determine changing workload and access patterns? How can we design methods that are robust to environmental changes?
- How does a client know whether the desired data is being pushed or has to be pulled?

7.2 THE CASE FOR DYNAMISM

Several factors can affect the performance of an information dissemination mechanism. One of these factors is the query arrival rate (or number of clients). Figure 7.2(a) shows a representative result on three information dissemination schemes:

- The *pure pull* scheme disseminates objects to clients based on their explicit requests.
- The *pure push* scheme disseminates objects by broadcasting them periodically as a flat broadcast program.
- The *static integrated* model (also known as the *hybrid* strategy) combines the two modes to disseminate frequently access data periodically and the

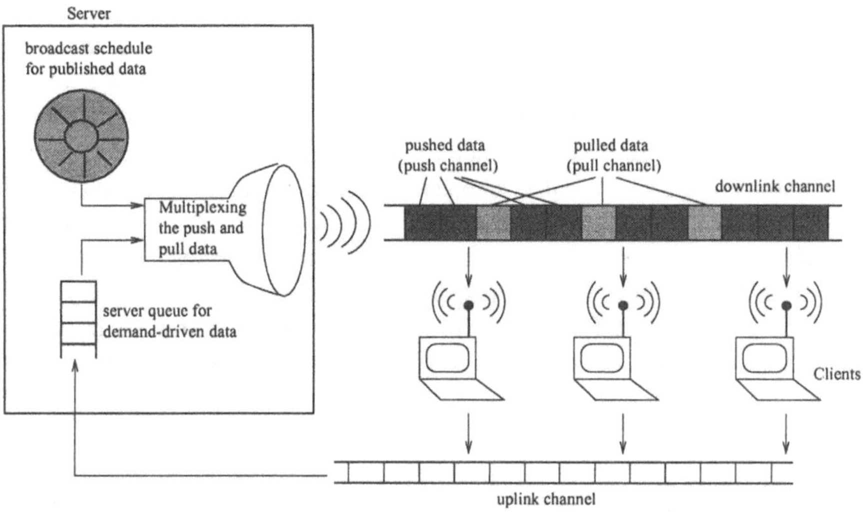

remaining less frequently accessed objects on demand. The scheme is static in the sense that the amount of data objects and the set of objects to be periodically broadcast, and the bandwidth allocated for these objects are predetermined. Moreover, there is no runtime mechanism to change them. Such a scheme has been proposed in [59, 121].

From the figure, we note that the pure push model has a large flat access time, Needless to say, this flat access time increases with the size of the database. The pure pull model, on the other hand, has a very low access time when the query arrival rate is low. This is because the number of requests is small and the pull channel is not congested. However, its performance deteriorates rapidly with increasing query arrival rates as the pull channel becomes congested. There is a certain query arrival rate, say u, whereby both the pure push and pure pull techniques perform equally well — when the query arrival rate is less than u, the pure pull model is superior; when the query arrival rate is greater than u, the pure push strategy performs best.

The same figure also shows two static hybrid strategies that combine both models. The hybrid-x% scheme allocates the top x% of the most popular objects to be periodically broadcast and the remaining (100-x)% to be broadcast on demand. Clearly, the static strategies do not perform that well enough for them to be attractive alternatives over the pure push and pure pull methods: they perform best only in moderate workload. For low query arrival rates, the average access time is dominated by the size of the pushed data (since the average access time on these objects are half the broadcast cycle). Clearly, when the query arrival rate is low, the pull channel is not congested, making the pure pull approach performs the best. On the contrary, when the query arrival rate is high, the pushed data may not be enough to meet the needs of all clients. This led to a large number of clients still having to request for data explicitly, rendering the pull channel becomes congested quickly. The resultant effect is that the average access time also increases rapidly. In this situation, the pure push scheme is superior. Finally, when the arrival rate can be nicely serviced by the pushed data without causing congestion on the pull channel, then the static approaches are superior. We note that, the two static schemes — Hybrid-10% and Hybrid-20% — have their own space of superiority. For low query arrival rate, Hybrid-10% outperforms Hybrid-20% since it has a smaller number of broadcast objects. On the other hand, Hybrid-20% outperforms Hybrid-10% for high query arrival rate.

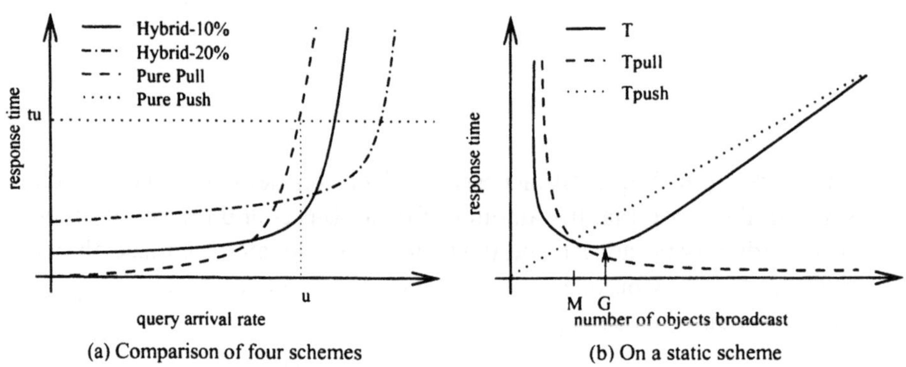

Figure 7.2. Case for dynamic hybrid scheme.

To further understand the inadequacy of a static scheme, Figure 7.2(b) shows the effect of varying the number of objects (to be periodically broadcast on the push channel) on the response time of a static strategy for a fixed query arrival rate. In the figure, T_{pull} and T_{push} denote the average response time of requests obtained from the pull channel and the push channel respectively. T is the average response time of all requests. As expected, the response time for requests directed at the push channel increases with the number of objects broadcast. When the number of objects broadcast is small (below certain threshold value, M), the server is unable to cope with the demand-driven requests, resulting in very high response time for queries that are served on the pull channel. However, as more objects are broadcast, the workload at the server becomes very small, and the response time decreases. The figure also shows the average performance of combining requests from both workloads (i.e., T). Clearly, it would be desirable to determine an optimal point (e.g., at G in the figure) to achieve the best overall performance.

From the above results, it is clear that though a static algorithm can perform best in some cases, it can also perform poorly. Furthermore, we note that the amount of data to be included in the push group can significantly affect the scheme. Other factors that may also affect a static scheme is the amount of channel bandwidth to be allocated for the push channel and the data skew. Since client access patterns and workload are expected to change, there is a clear need for dynamic strategies to be designed.

7.3 ADAPTIVE INTEGRATED MODELS

In this section, we shall present some of the existing work that integrate both the push and pull modes for information dissemination. Moreover, these schemes are adaptive in the sense that they are robust to the workload and/or access pattern changes. Before we look at the various schemes, let us first discuss how a client determines whether it should make an explicit request to pull the desired data object. The basic mechanism that has been adopted is for the server to broadcast some meta-data that indicate the content of the pushed channel. Clients must examine the meta-data to determine whether their desired objects are being broadcast in the pushed channel. If they are, then the objects can be obtained from the pushed channel; otherwise, clients must submit uplink requests. As such, objects that are pulled incur the additional latency of waiting for and examining the meta-data from the pushed channel.

7.3.1 Dynamic Scheduling Strategy

In [113], an integrated scheme known as *dynamic scheduling strategy* was proposed. The scheme is robust to workload (number of clients) but it assumes that the access frequencies of objects are known and that they do not change frequently. Data objects are organized into the two groups, the *publication group* and the *demand-driven group*. Objects in the publication group are periodically broadcast, while those in the demand-driven group are provided on demand. Let t_{bc} and t_{od} denote the average access times for the publication group and on-demand group respectively. The design of the scheduling strategy is based on the premise that it is not desirable to have $t_{bc} > t_{od}$ since the publication group contains the frequently accessed objects and is independent of the number of clients. Thus, the problem is reduced to determining the objects to be included in the publication group and the bandwidth to be allocated to the two groups such that $t_{bc} = t_{od}$ or $t_{bc} < t_{od}$.

Based on the premise, the following two observations were made:

- When the *access pattern* is uniform, i.e., every object is equally likely to be accessed, then the desirable relationship between the access time of the two groups should be $t_{bc} = t_{od}$. Moreover, the minimum average access time is obtained by either the pure pull or pure push technique. In other words, the performance of the integrated strategy will be bounded by the two techniques.

- When the access pattern is nonuniform, i.e., some objects are more popular, then the minimum average access time may be obtained by a hybrid strategy. Moreover, the preferred relationship between the access time of the two groups is $t_{bc} < t_{od}$ since popular objects are in the publication group.

Since under uniform access patterns, the best average access time is obtained by either the pure push or pure pull technique, the most straightforward adaptive scheme is to "switch" from one mode to the other based on some simple predetermined mechanisms. For example, if the server is currently employing a pure pull technique, then the server will swap to the pure push strategy when the system notices that the average access times for requests exceed the average access time for the publication group. Similarly, if the pure push technique is being employed, then when the number of clients[1] falls below a certain predetermined threshold, then the server will change to the pure pull approach.

When the access pattern is nonuniform, the integrated information dissemination strategy is more complex. The data objects are organized into a large number of *buckets*. Objects within a bucket have approximately the same access frequencies. The buckets are ranked according to the access frequencies such that bucket i has higher frequency than bucket j for $i < j$. Any addition or subtraction of data objects from the publication group is performed at *bucket-level*, i.e., all the objects within a bucket must be added or subtracted from the publication group.

The proposed dynamic integrated strategy is shown in Figure 7.3. The function *ComputeBroadcast(PG)* computes T_{bc}, the average access time for the publication group PG. The function *ComputeOndemand(OG)* computes T_{od}, the average access time for the demand-driven group OD. Note that T_{od} can be obtained from real processing of requests. Routines *Pull* and *Push* are for withdrawing data objects from and pumping data objects to the publication group respectively.

The framework begins (lines 1 and 2) by determining T_{bc} and T_{od} for the current workload. When the difference is beyond "tolerance" (line 3), one of the *Push* or *Pull* routines will be invoked. When the pull channel is underloaded, we can withdraw some data from the publication group (line 5). On the other hand, when the pull channel is overloaded, we can pump more data to the air (line 7). The *tolerance* is used to minimize "toggling" where data are pumped into/withdrawn from the air only to be withdrawn/pumped again, and so on.

Clearly, different heuristics can be employed for the push and pull routines. Figure 7.4 shows one such heuristic for routine *Push*. As shown, the heuristic first determines the average access time of the current workload, T_{avg}. Next, it explores the benefits of adding buckets, one at a time, from OG to PG. As

Input: Publication Group (PG), Demand-driven Group (OG) and tolerance
Output: PG, OG, bandwidthPG, bandwidthOG

$T_{bc} \leftarrow$ ComputeBroadcast(PG)
$T_{od} \leftarrow$ ComputeOndemand(OG)
if $|T_{bc} - T_{od}| > tolerance$ {
    if $T_{bc} > T_{od}$
        Pull(PG, OG, $tolerance$)
    else
        Push(PG, OG, $tolerance$)
}

Figure 7.3. The framework for the dynamic integrated model in [113].

each bucket is included, the bandwidth is reallocated between the new PG and new OG, and the average access time under the groups is recomputed. As long as adding buckets to the publication group can lead to a lower average access time, buckets will be added to the publication group. The best allocation will be remembered and returned. The *Pull* hueristic is essentially similar, and we chose to omit it here.

7.3.2 Adaptive Integrated Model

In the above scheme, the server is assumed to have information on the access patterns of the objects. Moreover, the access patterns are assumed to be fixed, or seldom change. Thus, it is only robust to workload changes and is suitable for static environments. Cai and Tan [27] presented an adaptive integrated model that is robust both to the workload and the access patterns. The approach identifies and isolates the bottleneck in order to tune the system for "optimal" performance.

Adaptiveness to Changing Workload

We shall first focus on how changing workload is handled. There are essentially two cases to handle: an increase in workload or a decrease in workload. To simplify our discussion, we only focus on dealing with an increase in the workload. Decreasing workload can be handled easily using a reverse approach to that of increasing workload.

When users' access patterns do not change, the user arrival rate can affect the performance of the integrated model. High user arrival rate means high demand rates on both the push and pull channels. Since the performance of the push channel is independent of the number of clients tuning to it, the bottleneck is therefore with the pull channel. More requests on the pull channel also imply

```
T_avg ← ComputerAvgTime(PG,OG,bandwidthPG,bandwidthOG)
bestPG ← PG
bestOG ← OG
bestbandwidthPG ← bandwidthPG
bestbandwidthPG ← bandwidthOG
bestAccess ← T_avg
gain ← TRUE
while OG is not empty AND gain is TRUE {
    bucket ← nextBucket(OG)
    PG ← PG ∪ bucket
    OG ← OG − bucket
    ReAllocateBandwidth(PG, OG, bandwidthPG, bandwidthOG)
    T_avg ← ComputerAvgTime(PG,OG,bandwidthPG,bandwidthOG)
    if T_avg < bestAccess {
        bestAccess ← T_avg
        bestPG ← PG
        bestOG ← OG
        bestbandwidthPG ← bandwidthPG
        bestbandwidthPG ← bandwidthOG
    } else {
        gain ← FALSE
    }
}
PG ← bestPG
OG ← bestOG
bandwidthPG ← bandwidthPG
bandwidthOG ← bandwidthOG
```

Figure 7.4. Procedure Push().

more requests on the uplink channel and more severe workload on the server. Since the server's processing capability can be improved easily by adding more and faster CPUs, the proposed solution addressed the other two factors: the uplink channel and the pull channel. There is an important difference in these two channels. The bandwidth of the uplink channel is usually fixed, while that of the pull channel can be adjusted (up to the maximum capacity of the downlink channel) by the server. Uplink channel congestion can only be eased by decreasing the demands on this channel, while the pull channel congestion can be eased by either redistributing some of the push channel bandwidth to this channel or decreasing the demands on this channel by shifting more pull data to the push channel. By observing the arrival rate on each channel, the server can determine which channel is the first to be congested and take the appropriate measure:

- Case 1: When the uplink channel is congested, to reduce the query arrival rate on this channel, the *only* measure is to increase the hit ratio on the push channel. This means shifting some (sub-)hot data from the pull channel to the push channel. If at this time, the pull channel is not congested, in order to reduce the access time on the push channel, the bandwidth on the push channel can be increased by decreasing the pull channel bandwidth.

- Case 2: If the pull channel is the first to be congested, there are two alternative solutions. The first is to redistribute some of the push channel bandwidth to the pull channel. Suppose δC is the bandwidth required for easing the congestion on the pull channel, then the average access time on the push channel is lengthened to (we note that the paper [27] adopted a flat broadcast program):

$$t_{push} = \frac{KD}{2(C_{push} - \delta C)} \tag{7.1}$$

  where K is the size of each data object, D is the number of data objects in the published group, and C_{push} is the amount of bandwidth allocated to the push channel. The second alternative is to shift more pull data to the push channel. Suppose another δD data in the pull group will be put on the push channel to ease the demands on pull channel. The effects on the average access time of the push channel is :

$$t_{push} = \frac{K(D + \delta D)}{2C_{push}} \tag{7.2}$$

  Thus the decision depends on which choice gains more. For effective performance, δC and δD have to be carefully determined. As such, a simple approach was employed: for adaptation of bandwidth, the bandwidth will be adjusted by increment of 1% of the total downlink bandwidth, and for adaptation of publication group size, the group size is varied in increment of 1% of the total number of objects. The process of adaptation is repeated until a satisfactory performance is achieved.

Adaptiveness to Changing Access Patterns

The second phenomenon that can affect the system performance is that of changing access patterns. In a wireless environment, since mobile clients move from cell to cell, it is expected that the access pattern will also change. As a result, some hot data may become cold, while cold data become hot. In this case, if the published data are not changed, the overall hit rate on the push channel will decrease, and more mobile clients will request data using the demand-driven mode. The latter may lead to congestion on the uplink and the pull channels. To avoid this problem, the cool data on the push channel should be shifted out and the warm data on the pull channel should be published. Since data in the pull group are provided on-demand, access statistics on these data are known to the server. From the statistics, the server knows what data in the pull group have become hot and can be promoted to the push group. However, it is difficult for the server to sense the access pattern change in the push group. In [27], feedback mechanisms were employed, i.e., clients submit feedback on the objects they have accessed to the server. Three factors have to be considered when collecting this feedback. First, since the uplink channel is a potential bottleneck and may already be heavily utilized when the access pattern changes, the feedback should not take up too much uplink bandwidth. Second, the process of feedback collection should be fast, so that the system performance can be restored quickly. Third, since mobile clients are usually resource restricted, the burden incurred by managing their own statistics should be minimized. Four methods were examined:

1. *Full PiggyBacking (FPB).* The clients maintain complete statistics locally about its access pattern and piggyback these statistics to the server with the next demand-driven request. We note that only statistics between piggy-backing stages need to be maintained. In this case, the server has a complete picture of the access patterns of clients who submit feedback. However, it only has partial picture of the overall access patterns since clients whose requests can be met by the push group need not submit their statistics to the server.

2. *Partial PiggyBacking (PPB).* This is similar to FPB, except that clients do not transmit all the statistics. Instead, the clients submit statistics in a *step-wise* manner, i.e., at each on-demand request, clients will submit a portion of the statistics. This can minimize the bandwidth requirement. However, the server only has a partial statistics of the clients each time.

3. *Full Probing (FP).* The server broadcasts a signal to all clients, and clients respond by directing all requests on the push data to the pull channel. In this way, the server can determine the access patterns of connected clients. Note that the server do not have the complete access patterns of clients.

4. *Partial Probing(PP).* This is the partial counter-part of the full probing scheme to minimize bandwidth requirement. Instead of directing all re-quests on the push data to the pull channel, the clients only direct a portion of them.

The benefit of probing-based feedback mechanisms is that clients do not need to maintain any statistics locally. However, when collecting the statistics, the traffic on the uplink channel will be increased significantly during the period of statistics collection, which may lead to uplink channel congestion. Piggybacking-based feedback mechanisms, on the other hand, require that each client maintains a small history of its access pattern. Whenever the client needs to submit an uplink request, the statistics can be piggybacked. This implies a longer query request. We note that the statistics is not submitted as long as clients' queries can be answered by the published data.

An Adaptive Mechanism

Figure 7.5 summarizes the above discussion with an algorithmic description of the adaptive dissemination mechanism. Essentially, the mechanism checks whether the cause of any performance degradation is due to access pattern changes (lines 2 and 9); otherwise, it would mean either uplink channel or pull channel congestion. For the former case, the data will be adapted from the published group to the demand-driven group accordingly based on the available statistics. For the latter case, the data and bandwidth adjustments are made as described above.

We would like to point out that the performance study in [27] showed that the piggybacking schemes perform better than the probing schemes.[2] The probing schemes require users to submit requests on the probed data to the uplink channel causing temporary congestion on both the uplink and pull channels. Moreover, it takes time for the system to restore the hit rate on the push channel. The piggybacking schemes, on the other hand, consume very little bandwidth for the purpose of statistics collection. PPB performs only marginally worse than FPB, making it a promising feedback mechanism to be employed.

7.3.3 Adaptive Scheme Based on Marginal Gains

In [101], an adaptive integrated scheme that is based on *marginal gains* was proposed. The approach does not make any assumption about the access patterns. The only assumption, which is not unreasonable, is that the server capacity is large enough to handle the requests on the *cold* set. The argument is that the cold set is expected to be very much larger than the *hot* set, and that objects that are to be broadcast repetitively should be restricted to hot set only.

Objects in the database are categorized into three different types using the concept of *temperature* which corresponds to its request rate λ. An object is said to be in the *vapor hot* state if it is very popular and hence should be placed in the publication group to be dissemiated under the push model. An object is in the *liquid warm* state if the number of requests received is moderate but not enough to justify their placement in the publication group, i.e., liquid objects are disseminated under the pull model. Finally, an object is considered *frigid (icy) cold* if it has not been requested for a while, and therefore, its temperature λ has practically dropped to 0 "degree". Requests for frigid objects are also

```
Input: Publication Group (PG), Demand-driven Group (OG)
Output: PG, OG, bandwidthPG, bandwidthOG

if congested(uplink_channel) {
    if accessPatternChange()
        adaptData(PG, OG)
    else { // workload change
        adjustData(PG, OG)
        adjustBandwidth(PG, OG, bandwidthPG, bandwidthOG)
    }
} else { // pull channel congestion
    if accessPatternChange()
        adaptData(PG, OG)
    else { // workload change
        // estimate cost according to Equation 7.1
        t1 → estimateCost1(PG, OG, bandwidthPG,bandwidthOG)
        // estimate cost according to Equation 7.2
        t2 → estimateCost2(PG, OG, bandwidthPG, bandwidthOG)
        if (t1 < t2)
            adjustBandwidth(PG, OG)
        else {
            adjustData(PG, OG)
            adjustBandwidth(PG, OG, bandwidthPG, bandwidthOG)
        }
    }
}
```

Figure 7.5. The adaptive integrated model in [27].

demand-driven. We note that only hot and liquid objects need to be maintained in memory, frigid objects (which form a large part of the database) can be stored on disk and accessed only when required.

Objects are disseminated using a *repetitive* non-flat broadcast program, i.e., some objects may be broadcast frequently, but the program is not periodic. This is realized as follows. Objects in the vapor state are maintained in a queue $\mathcal{V}$. The server picks the next object to broadcast from the head of $\mathcal{V}$. After the object is broadcast, the server removes it from the head, and moves it to the tail of $\mathcal{V}$. At the same time, the server reduces the object's temperature by some predetermined factor to reflect the cooling process of vapor data. This is necessary since there is no mechanism to detect the hits as well as misses on the object. At the end of each round of broadcasting all objects in $\mathcal{V}$, liquid objects that have vaporized are added to the tail of $\mathcal{V}$, while those vapor data whose temperatures have cooled to become liquid are removed from $\mathcal{V}$. Selective tuning is also supported by interleaving the list of data identifiers with the objects in $\mathcal{V}$.

Determining how each object's state may change is the crux of the scheduling strategy. Essentially, vapor data may cool down to liquid state, liquid data may be heated up to vaporize or cooled to freezing point, and frigid data may turn to liquid state before vaporizing. The change of state between liquid and frigid data is straightforward since the access rates of these data are known. To decide when vapor data should be "demoted" and liquid data be "promoted", Stathotos et. al. employed the concept of *marginal gain*. When data are removed (added) from the vapor queue, there is a drop (increase) in the average response time of pushed data and an increase (drop) in the average response time of pulled data. The marginal gains for the pushed data and pulled data correspond to the change in the response times of the pushed data and pulled data respectively. The average marginal gain δT in the expected response time is defined as the weighted average of the marginal gains for the data that are pushed and pulled. It was found that to operate at the optimal point may lead to very poor performance as a small change in the workload/access patterns can make the system unstable. To remedy this, the proposed approach operates at a sub-optimal point determined by a certain threshold value. Figure 7.6 summarizes the approach, which comprises three phases. In the first phase, vapor objects are demoted as long as their temperatures are lower than that of the hottest liquid object. In the second phase, if the expected response time is above the threshold value, more vapor objects are demoted (in the order of increasing temperature). Finally, in phase three, liquid objects may be promoted as long as the expected response time is below the threshold value. Note that it is possible for vapor objects that are demoted to vaporize immediately again.

7.3.4 Adaptive Scheme based on Adaptive Broadcast Protocol

Datta, et. al. [34, 35] adopted a slightly different approach to designing an adaptive scheme by constructing a *broadcast program online* prior to a broadcast epoch (i.e., the time at which a new broadcast is scheduled to begin). In some sense, the approach resembles scheduling strategies for pull-based broadcast systems in that the main challenge is to determine the objects to be broadcast in the next program. However, the approach differs in that it also determines the temporal duration that the objects need to be broadcast, once they are included.

The decision on which object to pick for inclusion in the next broadcast program is based on several novel concepts:

```
for each object in the vapor queue
    if its temperature is lower than that of the hottest liquid object
        demote it
v ←coldest vapor object
while the expected response time is above the threshold value
    demote v
    v ←coldest vapor object
l ←hottest liquid object
while the expected response time is below the threshold value
    promote l
    l ←hottest liquid object
```

Figure 7.6. Adaptive algorithm based on marginal gain [101].

- *Residence Latency.* The residence latency (*RL*) of a cell characterizes the average duration of a mobile user's stay in the cell. It can be computed *a priori* based on the average user movement patterns and cell geography. *RL* provides an indication on the duration in which an object would be broadcast for that cell. It was assumed that the size of a cell is such that the average query response time is much smaller than *RL*.

- *Popularity Factor.* The popularity factor (*PF*) of an object X at time T identifies the number of clients in the cell at time T who are interested in X. It is determined as follows. When a client requests for X, its PF_X value is increased by 1, and the server also notes the corresponding time. Let the timestamp of the ith increment to PF_X be at time T_X^i. Then, PF_X is decremented by 1 at time ($T_X^i + RL$), which reflects the (anticipated) departure of a client who requested X. The initial PF_X value is 0 (when there is no request for X).

- *Ignore Factor.* For an object X that has at least one request (*PF* value greater than 1), the ignore factor (*IF*) of X indicates the number of times X has been *ignored*, i.e., not picked for inclusion in the broadcast programs. The ignore factor of an object has a base value of 1. Note that *IF* value for X is bounded since it is no longer meaningful to consider the value once the client has left the cell.

Each object is assigned a priority value as follows:[3]

$$Priority = IF^k \times PF$$

where k is a factor that determines the relative importance of *IF* and *PF*. Intuitively, an object's priority increases with its popularity as well as the number of times it has been ignored. Such a measure favors popular objects while attempting to avoid *chronic starvation*. Avoidance of *chronic starvation* means that unpopular objects that have been requested should be included in some broadcast programs during the clients' stay in the cell.

In the *Constant Broadcast Size* (CBS) strategy, the program has a fixed size. Prior to a broadcast epoch, the server prioritizes all objects which have been requested, and adds objects to the broadcast program until the broadcast program size is reached. For all requested objects that are excluded from the program, their *IF* values are adjusted. On the other hand, the *Variable Broadcast Size* (VBS) strategy simply includes all objects with PF values greater than 0 in each broadcast program. Objects will remain in the broadcast for RL time units from their latest request, and are subsequently dropped from the broadcast (until they are requested again). Both strategies also interleave the data with indexes (using the $(1,m)$ index scheme [60]) to facilitate selective tuning.

An analytical study [34, 35] on the two methods suggested that the VBS strategy is superior at low system loads, but the CBS strategy is the better scheme at high system loads. This is expected since the VBS strategy may waste the downlink channel bandwidth by broadcasting an object multiple times in the duration corresponding to *RL* even though the demand for it may be low.

As noted, the content of each broadcast program is determined at the beginning of its broadcast. In addition, the index can only be built after the content of the broadcast is determined. This makes the scheduling overhead for the approach to be computationally expensive.

Notes

1 The paper [113] assumes that the server will have an estimate of the number of clients connected to it.

2 This is the first reported study on the relative performance of feedback mechanisms.

3 This concept is similar to the *priority index policies* of [103].

Chapter 8

SUPPORTING RELATIONAL OPERATIONS

Most of the work on information dissemination in wireless environments have largely focused on querying data objects. However, many applications employ databases, and queries involve relational operators such as selection, projection and join. For wireless computing to be truly relevant to many end-users, mechanisms to disseminate answer tuples in response to relational queries must be supported.

Database queries typically require manipulating a large portion of the base relations, and the query result sizes may consist of a large number of records. This implies that the communication channel can become congested very quickly. At the same time, query processing can be computationally expensive and evaluating complex queries at the client can significantly reduce the effective battery lifespan. We can also expect queries of a mobile client to exhibit a high degree of data locality. For example, some (canned) queries will be performed repeatedly. This can be exploited to design effective caching strategies. However, because of client's frequent disconnection, it is not easy to keep the cache content consistent with the relations at the server. Thus, it is a challenging task to support database applications in wireless environments.

In this chapter, we shall examine some of the work that have been done to facilitate and support relational operations. In Section 8.1, we shall look at cache coherency strategies under the pull model. Section 8.2 examines how join queries are evaluated under the push model.

8.1 CACHE COHERENCY STRATEGIES IN PULL-BASED SYSTEMS

Like the traditional client-server DBMS environment, portions of the database will be cached at the client to facilitate possible later reuse (e.g. for canned queries) to minimize future interaction with the server (to minimize contention on the wireless channel).

In [28], query results are treated as *views* and *materialized* at the client for future reuse. We shall refer to these materialized views as *client views*. More formally, suppose R and S are two base relations, then the following types of client views can be defined on the cached data:

select view:	$V_\sigma = \sigma(R)$
project view:	$V_\pi = \pi(R)$
join view:	$V_{\bowtie} = \bowtie(R, S)$

Initially, when a client first issues a query, the server constructs the client view, and transmits the data to the client. For a subsequent query that matches any of the views, the client can first check its validity and performs an update upon inconsistency. The basic scheme adopt the following basic strategy:

- The server broadcasts reports (that contain updates for wL time units) at intervals of L time units, and maintains the update logs for WL time units, where $W > w$.
- At the client, the client maintains the timestamp, T_c, at which its cache content is last valid.
- When the content of the client cache is invalidated (either partially or totally), the (query,T_c) pair will be submitted to the server. Note that this is different from traditional methods (see Chapter 6) where the list of objects or object ids is transmitted instead.
- Let T be the current time (this is the time when the server processes the client's query). If $T - T_c > WL$, then it implies that the client's cache content is totally invalid (i.e., nothing can be salvaged) and the query has to be processed from scratch. If $T - T_c < WL$, then the client view corresponding to the query can be incrementally reconstructed.

Several *incremental* cache coherency algorithms have been proposed in [28]. These schemes differ in two main components: the *invalidation/update of client view records*, and the *reconstruction of the client view*. Before we look at the various algorithms, we shall look at the information required to be maintained by the client and the server, as well as the content of the update report and the *differential file*.

8.1.1 Basic Information Maintained

For each client view maintained at the client, the client also keeps the following metadata to facilitate invalidation of the view [32]:

- the set of participating base relations, PR

- the set of attributes, ATT
- the set of conditions, COND
- the most recent timestamp that the view is valid, T_c

At the server, for each base relation R_i, incremental logs are maintained [88]. The incremental logs are partitioned into two logs: incremental deletion logs (D_i) and incremental insertion logs (I_i). Each record of D_i is a (RID, TS) pair where RID is the record id of a deleted record and TS is the timestamp at which the record was deleted. Each of I_i record's attributes are the attributes of R_i and one additional attribute for the timestamp TS. The values of the attributes of a record in I_i is the newly inserted values, and the timestamp TS represents the time at which the record is inserted. Modifications are modeled as a deletion followed by an insertion.

Note that the size of the update log cannot possibly increase indefinitely. As in previous work, the server only maintains the updates for up to WL time units (where L is the interval in which the update report is broadcast). As a result, if the client is disconnected for longer than WL time units, then its entire cache content has to be invalidated.

Next, let us look at the content of the update report. Two kinds of reports are supported. Under *update invalidation* scheme, the clients can only invalidate its views from the report. The report consists of a timestamp that indicates the time at which the report is generated, and a set of records with the following information:

- the relation identifier (R_i)
- the record identifiers (RIDs) of records updated during the interval $[T - wL, T]$
- for each RID value in the report, the timestamp that indicates when the record was last updated

The second type of report is based on the *update propagation* strategy where the updated records are broadcast in the report. Thus, clients can update their view from the content of the report. The content of the report is similar to that based on the update invalidation scheme except that additional attributes corresponding to those of the values of the entire records are needed of the updated records. For deletions, the corresponding values are set to NULL.

Finally, the updated results or logs have to be transmitted to the client. We shall refer to this as the *differential file*. The content of the differential file is the same as the update report under the *update propagation* strategy.

8.1.2 Scheme UI-C

Under UI-C, the update invalidation method is adopted and the client is responsible for reconstructing the view. The server periodically broadcasts, at intervals of L time units, the update history of the past wL time units. The server also maintains incremental logs for the past WL time units, where $W > w$. When a client reconnects and issues a query, it waits for the next invalidation report. Upon receiving the report, the client first compares the timestamp of the client view (corresponding to the query), say T_c, with that of report, say T. There are two possible cases:

1. If the disconnection time is short, i.e., $T - T_c \leq wL$, then the client will examine the records of the report that correspond to the participating relations of the client view (of the query). Any cached records that appear in the report are invalidated.

2. If the disconnection time is long, i.e., $T - T_c > wL$, then the client will be uncertain of the validity of the cache content.

When all the cached records are valid, the client view can be used to answer the query. However, if some records are invalidated or the client is uncertain of the validity of the cache, it will submit the pair (query, T_c) to the server. Since the client reconstructs the view, the task of the server is to transmit the necessary data to the client. There are three situations for the server to handle:

- If $T - T_c \leq wL$, the server will search the delete log and insert log for records (corresponding to the participating relations) that are updated after T_c. These records are used to create the differential file to be transmitted to the client. Note that for a join view, the entirety of both relations have to be transmitted. The client will then reconstruct the view based on the incremental view maintenance method in [88] as follows. Suppose that after several insertions and deletions, relation R becomes R' and S becomes S'. Let ΔL_R and ΔL_S be the incremental update log obtained from the logs of R and S respectively. Let $D_R(D_S)$ and $I_R(I_S)$ be the deletions and insertions found in ΔL_R (ΔL_S). Let V_{op} and V'_{op} denote the old and new views respectively, where *op* is one of σ, π and $\bowtie$. The views can be incrementally reconstructed using the following rules:

$$\begin{array}{ll}
\text{select view on } R\text{:} & V'_\sigma = V_\sigma - \sigma(D_R) \cup \sigma(I_R) \\
\text{project view on } R\text{:} & V'_\pi = V_\pi - \pi(D_R) \cup \pi(I_R) \\
\text{join view on } R \text{ and } S\text{:} & V'_\bowtie = V_\bowtie - V(D_R, *) - V(*, D_S) \\
 & \qquad \cup \bowtie (I_R, S') \cup \bowtie (R' - I_R, I_S)
\end{array}$$

  where $V(D_R, *)$ and $V(*, D_S)$ are records in $V_\bowtie$ for which $D_R \in R$ and $D_S \in S$. These rules reduce the computation of the new result (i.e., updated view) to applying the respective operator (select, project or join) on the update logs rather than the entire relation(s). Note that projection may require more work in order to handle duplicates, and that the computation of the updated join view requires the update logs as well as both the base relations.

- If $wL < T - T_c \leq WL$, the differential file is created in the same way as the above case. However, upon receiving the differential file, the client must perform the additional task of invalidating the records prior to reconstructing the view.
- If $T - T_c > WL$, then the server will transmit the base relations to the client. Upon receiving the relations, the client will process the query, and replace its existing client view with the new one.

8.1.3 Scheme UP-C

Scheme UP-C is very much similar to scheme UI-C except that the update propagation method is adopted. As a result, the only difference in processing from UI-C is when the disconnection time is short, i.e., $T - T_c \leq wL$, and so we shall only discuss this scenario here. When $T - T_c \leq wL$, the update report contains the values of updated records which can be used immediately by the client to refresh its cache content. We note that for joins, the base relations are required to be transmitted. Thus, the query can be answered using the updated client view without further communication.

8.1.4 Scheme UI-CS

In scheme UI-CS, the update invalidation method is adopted. Moreover, both the client and server *collaborate* to reconstruct the view. This scheme is different from UI-C in the reconstruction of the client view. It was observed that the consistency problem is caused by two actions: insertion and deletion (modification is treated as two consecutive actions: deletion followed by insertion). Let T_c denote the timestamp at which the client views are last valid, and D and I represent the records that are deleted and inserted respectively after T_c. The basic idea is to let the client reconstruct the view due to deletions, while the server reconstruct the view for insertions. The rationale is that for deletions, all the operations (selection, projection or join) can be performed on the client view (which is already at the client) by merely dropping those records in the view that are related to the deleted records. On the other hand, insertions require the inserted records and additional computation (for join in particular). To this end, the rules to reconstruct the views incrementally are revised to three logical parts as follow:

1. select view on R:

a) $V'_\sigma = V_\sigma - \sigma(D_R)$

b) I_R = **preprocess**(D_R, I_R)

c) $V'_\sigma = V_\sigma \cup \sigma(I_R)$

2. project view on R:

a) $V'_\pi = V_\pi - \pi(D_R)$

b) I_R = **preprocess**(D_R, I_R)

c) $V'_\pi = V_\pi \cup \pi(D_R)$

3. join view on R and S:

a) $V'_\bowtie = V_\bowtie - V(D_R, *) - V(*, D_S)$

b) I_R = **preprocess**(D_R, I_R); I_S = **preprocess**(D_S, I_S)

c) $V'_\bowtie = V_\bowtie \cup \bowtie (I_R, S') \cup \bowtie (R' - I_R, I_S)$

In other words, rebuilding work is split into two parts((a) and (c)): the rebuilding of the client view due to deletions to be performed by the client, and the rebuilding of the client view due to insertions to be performed by the server. The preprocessing step is used to handle insertions that occur at time $TI > T_c$ but are deleted at time $TD > TI$; otherwise, the server would have inserted records that are already deleted. The three situations that the server must handle are:

- If $T - T_c \leq wL$, the server will search the delete and insert logs for records (corresponding to the participating relations) that are updated after T_c. The ids of deleted records, together with the results due to the insertions are used to create the differential file to be transmitted to the client. In this way, for a join view, only the changes need to be transmitted instead of the entirety of both relations. Upon receiving the differential file, the client removes the deleted records from the client view before merging the inserted records of the client view.

- If $wL < T - T_c \leq WL$, the differential file is created in the same way as the above case. However, upon receiving the differential file, the client must perform the additional task of invalidating the records prior to merging the view.

- If $T - T_c > WL$, then the server will process the query and transmit the result to the client. Upon receiving the result, the client will replace its existing client view with the new one.

8.1.5 Scheme UP-CS

Scheme UP-CS is the counterpart of scheme UI-CS (just like UP-C to UI-C) that employs the update propagation method. As a result, the only difference in processing from UI-CS is when $T - T_c \leq wL$ where the update report can be used immediately by the client to update its cache content. Thus, the query can be answered using the updated client view without further communication. For joins, the server is also required to transmit the incremental join results.

8.1.6 Discussion

Extensive simulation studies were conducted in [28] to evaluate the performance of the four schemes. The results showed that update-propagation-based schemes are generally more inferior than update-invalidation-based schemes. This follows from the fact that update-propagation schemes incur larger bandwidth requirement and only benefit clients with short disconnection time. The result also suggested that UP-CS and UI-CS schemes perform better than UP-C and UI-C for small join view size (relative to the base relations). This is because a small join view size requires less downlink transmission. However, for large join view size, the client-based schemes are preferred for the same reasons (i.e., transmitting two smaller relations is cheaper than transmitting the large result). Though the client-based schemes are competitive to the client-server-based schemes, they require more powerful features in the mobile clients, and consume more energy. Thus, the client-server-based schemes are the more promising and preferred strategies.

8.2 JOIN PROCESSING IN PUSH-BASED SYSTEMS

Existing work on push-based broadcast systems have largely focused on organizing data to facilitate *selective tuning* for *simple* queries (e.g., single object retrievals) that involve a *single* relation [61, 62, 77, 111, 126]. Database queries, however, involve complex operations such as join, sort and aggregation. To support such operations effectively using the broadcast paradigm calls for more novel solutions. In this section, we examine join algorithms in a *push-based broadcast system* [105].

8.2.1 Broadcasting A Database

Current broadcasting approaches deal mainly with only a single relation. In a database environment, a database usually comprises several relations that are related in some ways. Given a database of k relations, a straightforward approach of adapting the existing solution (on single relation) is to broadcast the relations of the database one after another [99]. To facilitate selective tuning, each relation has its own index multiplexed with its records. For example, in Figure 8.1, the database comprises 3 relations, R_1, R_2 and R_3. Relation

R_1 has been split into three fragments (R_{11}, R_{12} and R_{13}), each with its own index fragments (I_{11}, I_{12} and I_{13}). Furthermore, each of the indexes has two additional entries: ($RelID$, $size$) pair to identify the relation (with id $RelID$) that the index is associated with and the size of the relation, and $k - 1$ triples ($RelID_j$, $offset_j$, $size_j$) for the other $k - 1$ relations (the jth relation is identified by $RelID_j, 1 \leq j \leq k-1$) to indicate the offsets to the beginning of these relations, and their sizes. Each version of a database with the associated indexes will constitute a *broadcast cycle*. For simplicity, we shall restrict our discussion to just two relations. Having more relations simply increase the access and tuning time in view of a longer broadcast cycle.

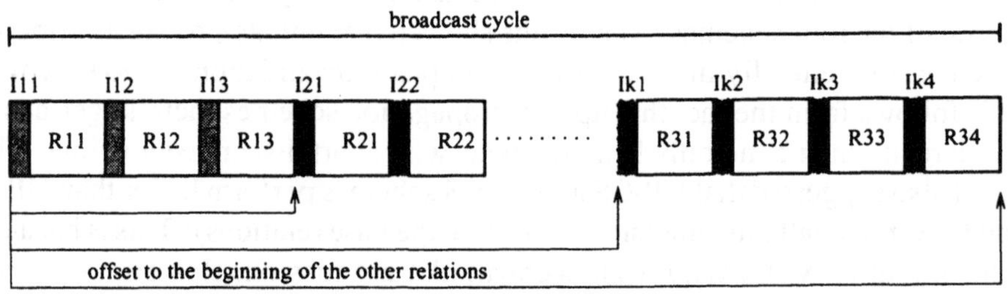

Figure 8.1. Broadcasting a database.

8.2.2 Join Processing

In this section, we look at several algorithms proposed in [105] for join processing between two relations R and S. R (S) has $||R||$ ($||S||$) records. Moreover, it was assumed that the mobile client has allocated M amount (in terms of number of records) of main memory and D amount (in terms of number of records) of disk space for the join processing.

Nested-Loops Join

To perform a join between relations R and S, a nested-loops join algorithm can be employed as follows:

1. The client tunes into the downlink channel, listens to an index, identifies the fragments of R that fit into the client's memory, and dozes off.

2. It wakes up when the fragments are broadcast, downloads the fragments, and dozes off.

3. It wakes up again when S is being broadcast. As each bucket of S is downloaded, the records are joined with those fragments of R in memory.

4. The above steps are repeated until all fragments of R have been examined. The number of iterations is given by $\lceil ||R||/M \rceil$.

Like the conventional nested-loops join in relational DBMS, the algorithm can be used to perform equi- and non-equi- joins, and the join condition can be on any attribute. However, as can be seen, the algorithm is not efficient in terms of both access time and tuning time:

- Its access time is at least $k \times (||R|| + ||S||)$, where k is the number of iterations. In fact, it fails to take advantage of the disk space that is available to reduce the access time.
- It is not energy efficient. Its tuning time to receive the data for join processing is at least $(||R|| + k \times ||S||)$.

Partition-based Join

In order to cut down on the tuning time, the relations can be partitioned to exploit traditional *partition-based* join algorithm [36]. In other words, relations R and S are partitioned on an attribute (called the *partitioned attribute*) into k buckets each so that only corresponding buckets of R and S need to be examined for join processing. While this method does not reduce the access time (since air is a sequential medium), the tuning time can be reduced to nearly $(||R|| + ||S||)$ for equi-joins on the partitioned attribute. As in the nested-loops join algorithm, this technique fails to exploit the disk space available in the mobile client.

We note that the partition-based algorithm can result in longer access time than the nested-loops join. This is because in the nested-loops join, the number of records to be kept in the memory is exactly (or never more than) the size of the memory. But, for the partition-based join, buckets may not fill up the entire memory space. As a result, it is possible that the number of iterations to complete the processing be more than that required by the nested-loops join. On the other hand, for a R bucket that does not fit into the memory space, the corresponding S bucket has to be scanned multiple times, each on a small portion of the R bucket (like the nested-loops join).

Disk-based Partition Join

The partition-based join can be enhanced to reduce the access time by using the disk space in the mobile client. To illustrate the idea, let us consider the example shown in Figure 8.2. In this example, we did not show the indexes. Here we have two relations, R and S, that have been partitioned into 10 buckets each. Suppose $R_1 + R_2 \leq M$ but $R_1 + R_2 + R_3 > M$, $R_3 + R_4 \leq M$, $R_3 + R_4 + R_5 > M$, $R_5 \leq M$ and $R_5 < D$ but $R_5 + R_6 > M$. Now, during the first iteration, the processing involves 5 steps:

1. Download R_1 and R_2 into the memory of the client. When an index exists, the client needs to stay active only during the index traversal and downloading.

2. Download R_5 into the mobile client and store it on the disk. When an index exists, the client needs to stay active only during the index traversal and downloading.

3. Download S_1 and S_2 to join with R_1 and R_2 in memory.

4. Read R_5 from disk into main memory. This steps occurs while S_3 and S_4 are being broadcast.

5. Download S_5 to join with R_5.

First, we note that in step 2, we have downloaded R_5 instead of R_3 and R_4. This is because storing R_3 and R_4 in the disk is not going to help in reducing the access time, as S_3 and S_4 would have been missed by the client by the time R_3 and R_4 are loaded into the memory from the disk. Thus, there is a need for a "gap" between buckets of R.

Second, the timing of step 4 is crucial. R_5 must be in memory within the span of time that S_3 and S_4 are being broadcast, and before S_5 is being broadcast. Otherwise, we will miss S_5 and the join in step 5 cannot be performed until the next cycle. We also note that the number of buckets to "skip" (S_3 and S_4 in this case) depends on how fast the data is being read into memory (i.e., processor and disk speed) and the bandwidth of the broadcast channel.

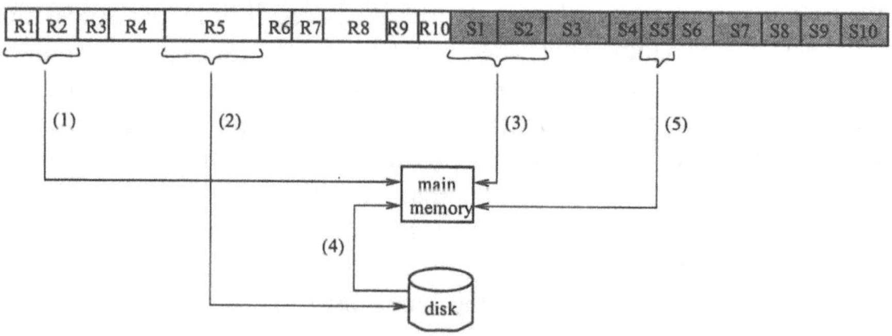

Figure 8.2. An example to illustrate the use of disk space to reduce access time.

Intuitively, such an approach can reduce the access time to $k \times (\|R\| + \|S\|$ where $k = \lceil \|R\|/(M + D) \rceil$. This can be significant especially for large D. The tradeoff is that the tuning time is expected to be higher since disk accesses are high energy consuming operations.

The problem essentially becomes one of finding an optimal schedule that reflects how buckets should be joined. Figure 8.3 shows the algorithmic description of a *greedy algorithm* that generates such a schedule. The algorithm comprises several iterations. Each iteration corresponds to one pass over the relations. In each iteration, several buckets of R will be selected to be staged in memory (lines 4-14) and several buckets will be selected to be loaded into the disk (lines 17-37). These buckets will participate in the join process within one broadcast of S. This process is repeated until the entirety of R has been selected. We note the following:

- First, since the memory size is M, the disk space D can be organized as a set of smaller regions whose size is M (except possibly for the last region). Where D is smaller than M, then there is only one region of size D. In this way, buckets are stored in the disk to fit the region size since regions are the units to be staged in memory.

- Second, as illustrated in the above example (see Figure 8.2), because buckets of a region (i.e., partitions of R) have to be staged while S is being broadcast, regions should be separated by some number of buckets. This "gap" is determined by S since it is S's buckets that are being broadcast.

The resultant schedule is partially built in each iteration. We note that it is sufficient to include in the schedule the buckets of R (since the join is performed with the corresponding buckets of S). The join processing can then be performed based on the schedule determined.

8.2.3 Supporting Joins on Unpartitioned Attribute

Partition-based joins are effective for joins that are based on the partitioned attributes. However, very often, joins may also be performed on other attributes. While nested-loops algorithm can be employed in such cases, Tan [105] proposed the use of bit filters to better utilize the scarce memory resource of the mobile client. Consider just one such non-partitioned attribute used for joins. The approach works as follows. Each record of relation R (and S) is tagged with a bit. The bit is set if there is a corresponding record in S (R) with the same join attribute value. In this way, as records of R are scanned, only those that will contribute to the join result will be retained in the main memory. This will reduce the number of iterations in a nested-loops join and hence the access time for join processing. Clearly, the scheme is not restricted to just one bit per record. In fact, for every non-partitioned attribute that are used frequently for join processing, this approach can be applied.

Note that since this approach is essentially a nested-loops algorithm, the disk-based partition join algorithm behaves in the same manner as the partition-based join algorithm. This is because the disk space is not utilized at all as it does not provide additional benefits.

8.2.4 Broadcast Organization to Support Selective Tuning

To support selective tuning in the proposed join methods, the broadcast is organized as follows. Each relation is broadcast together with a distributed

```
Input: MemorySize = M, DiskSize = D, n buckets of R and S
Output: Schedule
i ← 1 // ith iteration
schedule ← ∅
while some buckets have yet to be joined {
        // memory loading phase
        j ← 0 // jth bucket
        memoryOK ← TRUE
        P_Ri ← ∅
        while j ≤ n & memoryOK {
                j ← j + 1
                if R_j has not been considered {
                        memoryOK ← checkMemory(P_Ri, R_j, M)
                        if memoryOK {
                                mark R_j as considered
                                P_Ri ← P_Ri ∪ j
        }}}
        diskSpaceOK ← TRUE
        c_stage ← stageCost(M) // M is used since at most M records can be staged
        while diskSpaceOK & j ≤ n {
                // by pass some buckets of S
                P_S ← j
                c_byPass ← byPassCost(S, P_S)
                while j ≤ n & c_byPass < c_stage {
                        j ← j + 1
                        P_S ← P_S ∪ j
                        c_byPass ← byPassCost(S, P_S)
                }
                // disk loading phase
                diskSpaceMemory ← TRUE
                P_R ← ∅
                while j ≤ n & diskSpaceMemory {
                        j ← j + 1
                        if R_j has not been considered {
                                diskSpaceMemory ← checkDisk(P_R, R_j, M, D)
                                if diskSpaceMemory {
                                        mark R_j as considered
                                        P_R ← P_R ∪ j
                }}}
                P_Ri ← P_Ri ∪ P_R
                diskSpaceOK ← diskOK(P_Ri, R_j, D)
        }
        i ← i + 1
        schedule ← schedule ∪ P_Ri
}
```

Figure 8.3. The greedy scheduling algorithm.

index that facilitates single record retrievals. This is similar to that discussed in Section 8.2.1. In addition, for every pair of relations to be joined, a *join index* is also broadcast at the beginning of the relations. Entries of the *join index* contain bucket identifiers of the pair of matching buckets and their offsets in the air storage, and their sizes. For join processing, the join index will be accessed first. The information in the join index can then be used to determine the schedule for the disk-based partition join algorithm.

8.2.5 Discussion

In [105], a simulation study was conducted to evaluate the effectiveness of the three join algorithms on both the access and tuning time: nested-loops join (NLoop), the partition-based join (Pmem), and the disk-based partition join (Pdisk). We note that the tuning time includes the energy consumed to read the relations as well as the energy consumed for the I/O operations. It is assumed that the energy consumed by reading/writing data from/to the disk is 4 times more than that of reading data from the air.

Figure 8.4 shows the effect of the ratio between the wireless and disk bandwidth, ω, when the join is on the partitioned attribute. ω is a critical factor for the disk-based partition join. A small value for ω would be favorable for Pdisk as fewer buckets of S need to be "skipped" when reloading the memory with data from disk. A large ω value would thus mean smaller gain (in terms of access time) from exploiting the disk. This study varies ω from 0.2 to 6. The result shown in Figure 8.4 confirms the above intuition for access time. In fact, it shows that even if the wireless bandwidth is 6 times more than that of the disk (which is high considering wireless bandwidth are fairly low), the disk-based partition join remains very effective in reducing the access time. The result also shows that the tuning time is largely unaffected by ω for all algorithms. Moreover, it is interesting to note that the tuning time for Pdisk is moderate (about twice as much as Pmem, but 20% of NLoop) making it a promising strategy to adopt.

Figure 8.5 shows the result on the effectiveness of the algorithms for non-partitioned attributes. Here, the percentage of R records with matching S records is varied from 20% to 100%. From the figure, we clearly see that by employing the concept of bit filtering, the performance of the partition-based join and disk-based partition join (in terms of both access and tuning time) can be drastically improved, especially when the percentage of R records with matching S records is low. This follows from the fact that the memory is better utilized and hence the number of cycles to complete the join processing can be reduced.

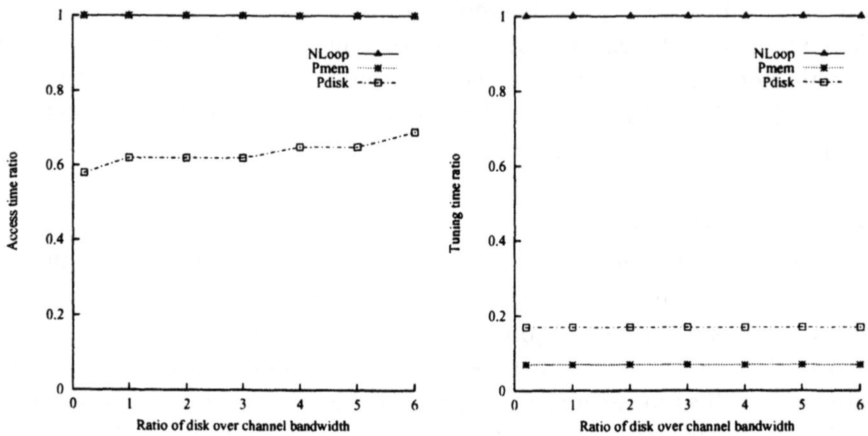

(a) Access time (b) Tuning time

Figure 8.4. Effects of ω

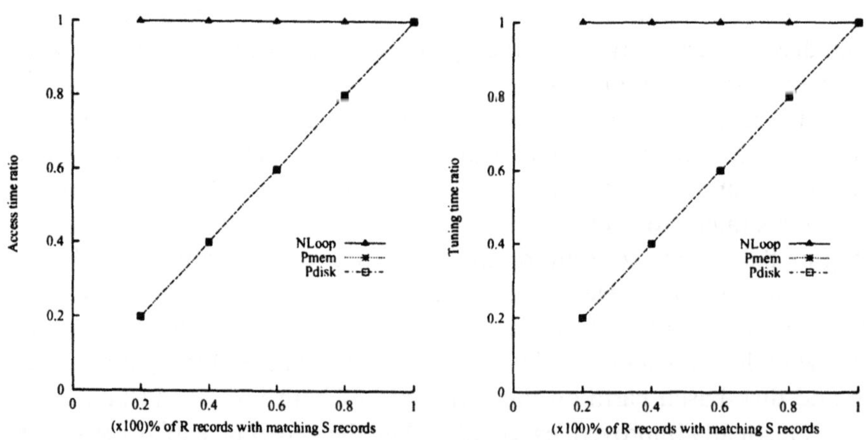

(a) Access time (b) Tuning time

Figure 8.5. Effects of % of R records with matching S records.

Chapter 9

CONCLUSIONS

In this book, we have examined information dissemination mechanisms in wireless computing environments. This chapter summarizes the main issues and approaches that have been adopted in the literature. There are many interesting topics that we believe are worthy of further exploration. We also provide suggestions on some of these in this chapter.

9.1 SUMMARY

In this book, we have looked at the issues and solutions of wireless information dissemination mechanisms. Most of these mechanisms are designed to deal with the main constraints in wireless computing environments: the error-prone channel and its limited bandwidth, the short battery lifespan of battery-operated portable devices and frequent disconnection of mobile clients.

Broadcast Scheduling. Since the wireless channel is a sequential medium, disseminating information on the wireless channel calls for the data to be effectively organized for broadcasting. In the push model, the challenge lies in the organization of the broadcast data to keep the access time low. Existing solutions typically organized the broadcast data into a non-flat broadcast program based on *priori* knowledge of the access probabilities of each object. In a non-flat program, objects that are frequently demanded are broadcast more often than those that are less popular. Such programs provide short access time to popular objects at the expense of long access time to unpopular objects.

For the pull-based model, the wireless channel can become a bottleneck or be poorly utilized especially under high workload. The solutions in the literature essentially deploy scheduling strategies at the server to pick the most appropriate object to be disseminated next based on the pending requests.

Index on Air. Energy, being a limited and precious resource at the client, needs to be conserved. Clients need to operate in the CPU (active) mode to listen to the broadcast data. This, unfortunately, is an energy consuming operation. The approach that has been adopted to reduce this energy consumption is to interleave the broadcast data with meta-data (i.e., indexes). The index allows clients to selectively tune to the relevant portion of the broadcast. Thus, clients can operate in the less energy consuming doze mode most of the time and stay in the active mode only when the objects relevant to them are being disseminated. Indexing of both flat and non-flat programs have been studied.

Reliable Broadcast. Wireless communication is error-prone. Data may be corrupted or loss. In a wireless environment, the consequence of this means waiting for the next copy of the data to be disseminated. This can increase both the access time and energy consumption. To deal with the unreliable wireless networks, additional meta-data are injected into the broadcast program. In the presence of errors, such meta-data allows the clients to continue to listen to the current broadcast program for its objects (unless the desired objects have been missed or corrupted) instead of waiting for the next broadcast program. Existing work are focused on dealing with failures in accessing the indexes. This is because a corrupted index bucket can invalidate the processing that has been performed, and may require restarting the entire retrieval process.

Cache Invalidation. Caching can reduce the channel contention as well as conserving energy and money. However, because clients frequently disconnect (to conserve energy), it becomes difficult for them to determine if their cache content are still valid after they have resumed connection. Existing work adopted an approach that disseminates update report to clients periodically. The update report contains information on recently updated objects, which a connected client can use to invalidate its cache content. We have examined several techniques in this book, some of which are energy-efficient schemes.

Integrated Push/Pull Model. It is a well known fact that neither the push model nor the pull model is superior under all circumstances. In particular, the push model is independent of the number of clients listening to the broadcast data; the pull model, on the other hand, has better access time performance when the system load is light. The main issue is to design an integrated model that can perform well in most, if not all, cases and should be robust to workload and/or access pattern changes. The difficulty lies in gathering statistics on the clients access patterns. In existing solutions, the server predicts the access patterns of clients based on partial statistics gathered from clients. These statistics are obtained either by clients who submit feedback on their access patterns or through clients' demand-driven requests. Based on the statistics, the server adaptively balances the demand on the two models to optimize the overall system performance.

Relational Operations. Unlike single object retrieval, relational operations such as join typically generate a large number of answer tuples. This can result in channel congestion very quickly. As such, existing solutions have adopted a broadcast-based approach. In other words, the relations are broadcast, and clients will access these relations and perform the operations locally. We have presented join processing algorithms in this book.

Clients queries are also expected to exhibit a high degree of data locality. Thus, clients are expected to cache their query results for reuse. As in single object retrieval scenario, there is a need to maintain the cache consistent with the relations at the server. In the literature, incremental cache coherency schemes have been adopted. These schemes are based on two main steps: invalidation of client cache content based on update reports that are periodically broadcast, and reconstruction of the client cache using an incremental strategy.

9.2 RESEARCH DIRECTIONS

We are only beginning to understand the basic issues in designing dissemination mechanisms for wireless computing. The following are some topics that we believe will draw much research attention.

Dynamic Broadcast and Consistency. In our preceding discussions, we have focused on the issues of generating broadcast programs, and have ignored the issues that arise as a result of changes to the broadcast programs. However, in practice, the broadcast program must accommodate changes. There are three types of changes [2]. First, updates may result in new objects being added to as well as existing objects being removed from the program. Second, the values of the objects in the program may change as a result of updates. Finally, for programs that are periodic, the programs themselves may change due to changes in access patterns and workload.

Changes to the broadcast program raise consistency issues for push-based systems. One straightforward solution is to allow clients to tune in only at the beginning of the broadcast program since the broadcast program can be considered a consistent unit. This, however, means longer access time, especially if one should just miss the beginning of the broadcast program. Instead, updates of a broadcast program can be included in subsequent broadcast programs so that clients only need to listen to *dirty* buckets that are relevant to them. There are some work done in this direction [4, 14, 35, 49, 81, 82] and more effort is certainly needed.

Moreover, traditional consistency model that is based on *serializability theory* may no longer be appropriate. In [4], several semantic models have been examined. For example, allowing clients to read only the *latest value* from a broadcast program is a weaker model than the serializability model since there is no concept of transactions. For clients who cache objects, an appropriate consistency model is *quasi caching* [9]. Under this model, a cached object whose value is within an allowable deviation (specified by each client's coherency condition) can be used to answer a query on the object. Efforts in addressing this issue are forthcoming [95] but more extensive study is needed to determine an appropriate consistency model for wireless computing.

Broadcast Scheduling. Previous work on broadcast scheduling has assumed that the rate at which an object is broadcast is dependent on the demand for that object. Such a strategy is "optimal" in the sense that it minimizes the average access time amongst all clients over all objects. In a real-time database environment [87], minimizing the average access time ceases to be the main performance criterion. Instead, guaranteeing that timing constraints imposed on data retrieval will be met becomes the overriding concern.

Some recent work have begun to address this problem. In [17, 65], the push-based approach is adopted. There, it is assumed that the server knows a priori the timing constraints of each object. Since the actual timing constraints may differ from that used by the server in determining the broadcast program, the approach is considered *probabilistic*. [43, 124], on the other hand, adopted an integrated model instead. The channel bandwidth is logically split into two, one for disseminating hot objects periodically, the other for on-demand requests. A query object is scheduled to be broadcast on the pull channel if the object is not an hot object or if it is in the hot set but its deadline cannot be met by the periodic broadcast. The *earliest deadline first* heuristic is adapted to work in this environment. As time critical applications are expected to become more common, this area will be a fruitful one to work on.

Another problem that is worth further investigation is the estimation of clients access patterns. As noted, most of the existing work on generating broadcast program for push-based context assume that the access frequencies is known. This is not very realistic in a mobile computing environment where clients move from cell and cell and many clients enter and leave a particular cell. Obtaining feedback from clients can be expensive as energy has to be expended to transmit the feedback. Moreover, the uplink channel can become a bottleneck. A promising method that is based on statistical inferences has been proposed recently [125]. The idea is to estimate hits on the push channel based on the requests submitted by clients. We believe that employing statistical methods is a right direction, and more extensive research in this direction is needed.

Reliable and Secure Broadcast. As mobile computers are likely to play an important role in future large-scale distributed systems, such as navigational and military applications, designing reliable and secure broadcast becomes more critical. In [21], the broadcast disks scheme was adapted to generate a fault-tolerant and secure program based on the *information dispersal algorithm* [85]. The scheme allows the system to adjust the amount of redundancy used in the transmission. The issues of optimizing the program and selective tuning have not been addressed. Moreover, more work are needed to study the effect of bursty errors, encryption/authentication protocols in push-based model, and how to further boost the security/privacy of push-based approach.

Joint Broadcast with Caching and Prefetching. In push-based dissemination systems, the concept of caching and prefetching has to be reexamined. Traditionally, clients typically retain data that are most frequently used. This is a reasonable approach since the cost of a cache miss is "constant" for all objects. However, in push-based models, the cost of a cache miss by clients depends on when the object will appear next in the broadcast. Intuitively, we should cache objects that are frequently accessed as well as objects that are not frequently broadcast. The former can reduce cache miss while the latter can minimize the cost of cache miss. Some work have recently been done to design schemes that balance the two factors [2, 3, 102]. Similarly, novel prefetching strategy has to be designed to reduce the cost of servicing a miss [4]. More recently, Hu et. al. proposed combining caching with an integrated model [54]. Because of the benefits of combined models, this area certainly deserves further investigation.

Exploiting Multiple Channels. The work reported in this book have restricted to single downlink channel. One can also conceive an environment in which the server can disseminate information on multiple downlink channels, and clients can subscribe to as many channels as they want (and can afford). Much effort has been expended to generate broadcast programs for single channel environments. However, research on multiple channels environments are only beginning to surface. Research in this area is expected to be fertile and relevant, especially in the following topics:

Generation of Broadcast Program. To exploit multiple channels, the same program can be broadcast on multiple (or all) channels such that the program in each channel is "staggered", i.e., the program on channel i is delayed $(i-1)\tau$ time units as compared to the first channel. Thus, if an instance of a data object appears at time t on channel j, then the corresponding instance of the same object will appear at time $t+\tau$ on channel $j+1$. In this way, a client who tunes to more than one channel can receive its desired data object in a shorter access time compared to the single channel case: if the client misses an object in one channel, it can tune to another channel instead of waiting for the same object in the next broadcast of the same channel. The main issue in generating the broadcast programs for multiple channel is to identify how the programs should be staggered on multiple channels to minimize the mean access time. Vaidya and Hameed [117, 51] have proposed broadcast programs on multiple channels to reduce the access time. The techniques are adaptations of the single channel programs. Simulation and analytical studies confirmed that the access time can be significantly reduced. To our knowledge, there is no reported work that explores different broadcast programs on different channels.

Facilitation of Reliable Information Dissemination. Since wireless environment is error-prone, multiple channels can help to facilitate reliable transmission. This is because data can be replicated across multiple (or all) channels, and if some channels are unreliable, the desired objects can be obtained from other channels. Leong and Si [78] studied the replication of broadcast programs on all channels in providing a reliable wireless environment for single object and file retrieval, and showed its feasibility and practicality.

Selective Tuning. To our knowledge, there is also no reported work on exploiting selective tuning in multiple channels context. One approach is to allocate some channels for indexes. These indexes are more complicated than those designed for single channel environments because they may have to include information on all instances of an object in all channels in which in the data is broadcast. Another approach is to replicate index-based broadcast programs on multiple channels. This is simple, but it will require the client to tune to the index for all channels that it tunes to. How effective are each of these methods or novel methods in term of tuning time require further study.

Allocation/Optimization of Channel Bandwidth. Given that we have multiple channels, the allocation of channels to support push and pull models, and the indexes become an issue, i.e., how many channels should be allocated to disseminate data by the push model, and how many should be allocated for the pull model, and how many should be allocated for meta-data (such as indexes). Optimizing the channels allocation remains largely an unexplored field. To our knowledge, there is only one reported work [76]. Even with regards to channels that are allocated to the push model, we can also simulate the concept of broadcast disks by allocating different "disk" to each channel. Popular data can be placed on more channels so that they have a larger overall bandwidth, while unpopular data objects can be given limited bandwidth by disseminating them on a single channel. Optimizing such allocation has not been addressed in the literature.

Cache Invalidation Strategies. Most of the research on cache invalidation strategies disseminate invalidation reports periodically. In addition, the content of the report reflects only the latest change for objects which have updates in the last wL time units, where w is the update broadcast window and L is the broadcast interval between the dissemination of reports. The effectiveness of such schemes depend on the window size, w. The larger the window, the more "immune" the client is to disconnection, but the overhead of additional bandwidth requirement can be significant.

Given a fixed window size, w, there are several limitations in these schemes. First, suppose an object o_1 is updated at time T_0, after which it is never updated. Then, it will not be included in any invalidation report after the timestamp exceeds $T_0 + wL$. If a client caches o_1 at $T_1, T_0 < T_1 < T_0 + wL$, disconnects immediately, and reconnects after $T_1 + wL$, it will have to submit an uplink request upon the next query (even though o_1 is still valid!). The problem is exacerbated when o_1 is a popular object accessed by many frequently disconnected clients, which may lead to lots of unnecessary uplink traffic. For such objects, it may be beneficial to include them in the invalidation report as well. We also observe that for frequently updated objects, even if a client is connected, it must still invalidate the object and submit an uplink request. Thus, it may not benefit to include them in the invalidation report.

Another problem with existing approaches concerns the size of the invalidation report, whose demand on the downlink bandwidth means a reduction on the effective bandwidth used for disseminating answer objects. For example, it was shown in [106] that a simple strategy that queries the server on the validity of the cache content can outperform the bit-sequence scheme because of its large invalidation report size.

Therefore, one promising direction for future work is to design adaptive mechanisms to facilitate cache invalidation. Such mechanisms essentially attempt to promote *collaborative cache invalidation* where the clients perform some invalidation based on the invalidation report while the server evaluates the rest. In this way, only relevant information need to be disseminated in the invalidation report to maximize its effectiveness, and minimize bandwidth requirement. For example, the window size of the report need not be restricted to objects that are updated within the last wL time units. Instead, the window size can vary from reports to reports and objects to objects. As we have discussed, popular objects that are not frequently updated can be included in the report, i.e., the pair $[id_j, ts_j]$ reflects that object j with id_j is updated at ts_j, where $ts_j < T - wL$ and T is the current time. We note that not all objects whose last update timestamps are smaller than $T - wL$ are to be included in the report (otherwise, we end up disseminating all objects). Similarly, frequently updated objects can be excluded even if their update timestamps are very close to the timestamp of the current report. The former will demand more channel bandwidth, while the latter will reduce the channel bandwidth for the report. Methods are needed to identify the optimal set of objects to be included so that the bandwidth requirement can be minimized. In [16], two adaptive methods based on the Broadcast Timestamp schemes were reported. There, each object has its own window size which are updated based on the demand on the object and the rate at which it is updated. Information on the objects are then broadcast based on their respective window sizes. We believe that other cache invalidation methods will also benefit from such dynamic adjustments. There are still many outstanding issues to be resolved: how should the log window for each object be maintained? how much information do we need to design an optimal solution? what about the complexity and performance of these schemes? etc. This area will be fruitful and worthy of further research.

Relational Operations. We are only beginning to see research activities on supporting database operations in wireless environments. This area will be very fruitful and challenging. First, support should be provided for other operations such as sort and aggregates. Second, instead of broadcasting the entirety of a relation, techniques to reduce the bandwidth utilization should be explored. For example, only the frequently accessed portion of a relation needs to be broadcast, while the remaining portion can be provided on demand. This, however, calls for more complex join algorithms to be designed. Third, the tradeoffs between client-based, server-based and client-server-based schemes need a more extensive investigation.

Testbed and Benchmarks. Most of the reported work in the literature have employed simulation in their performance study. While simulation models can provide insights to performance of a scheme, there are typically inherent assumptions that are not realistic and they cannot adequately capture all parameters. Thus, it is necessary to develop a testbed and a set of benchmarks.

References

[1] A. Acharya and B. Badrinath. Checkpointing distributed applications on mobile computers. In *Proceedings of the 3rd International Conference on Parallel and Distributed Information Systems*, pages 73–80, Austin, Texas, September 1994.

[2] S. Acharya, R. Alonso, M. Franklin, and S. Zdonik. Broadcast disks: Data management for asymmetric communication environments. In *Proceedings of the 1995 ACM-SIGMOD International Conference on Management of Data*, pages 199–210, June 1995.

[3] S. Acharya, M. Franklin, and S. Zdonik. Dissemination-based data delivery using broadcast disks. *IEEE Personal Communications*, 2(6), December 1995.

[4] S. Acharya, M. Franklin, and S. Zdonik. Disseminating updates on broadcast disks. In *Proceedings of the 22nd International Conference on Very Large Data Bases*, pages 354–365, Bombay, India, September 1996.

[5] S. Acharya, M. Franklin, and S. Zdonik. Balancing push and pull for data broadcast. In *Proceedings of the 1997 ACM-SIGMOD International Conference on Management of Data*, pages 183–194, May 1997.

[6] S. Acharya and S. Muthukrishnan. Scheduling on-demand broadcasts: New metrics and algorithms. In *Proceedings of the Fourth International Conference on Mobile Computing and Networking*, pages 43–54, Dallas, TX, October 1998.

[7] D. Aksoy and M. Franklin. Scheduling for large-scale on-demand data broadcasting. In *Proceedings of the 1998 IEEE INFOCOM Conference*, San Francisco, CA, March 1998.

[8] I.F. Akyildiz and J.S.M. Ho. Dynamic moible user location update for wireless pcs networks. *Wireless Neworks*, 1(2), 1995.

[9] R. Alonso, D. Barbara, and H. Garcia-Molina. Data caching issues in an information retrieval system. *ACM Transactions on Database Systems*, 15(3):359–384, September 1990.

[10] R. Alonso and S. Ganguly. Query optimization for energy efficiency in mobile environment. In *Proceedings of the 1993 Workshop on Optimization in Databases*, Aigen, Austria, September 1993.

[11] R. Alonso and H. Korth. Database systems in nomadic computing. In *Proceedings of the 1993 ACM-SIGMOD International Conference on Management of Data*, pages 388–392, June 1993.

[12] M.H. Ammar and J.W. Wong. The design of teletext broadcast cycles. *Performance Evaluation*, 5(4):235–242, 1985.

[13] K.R. Baker. *Introduction to Sequencing and Scheduling*. Wiley, 1974.

[14] D. Barbara. Certification reports: Supporting transactions in wireless systems. In *Proceedings of the 17th IEEE International Conference on Distributed Computing Systems*, May 1997.

[15] D. Barbara and T. Imielinski. Sleepers and workaholics: Caching in mobile distributed environments. In *Proceedings of the 1994 ACM-SIGMOD International Conference on Management of Data*, pages 1–12, June 1994.

[16] D. Barbara and T. Imielinski. Sleepers and workaholics: Caching in mobile distributed environments. *VLDB Journal*, 4(4):567–602, 1995.

[17] S. Baruah and A. Bestavros. Pinwheel scheduling for fault tolerant broadcast disks in real-time database systems. In *Proceedings of the 13th International Conference on Data Engineering*, pages 543–551, April 1997.

[18] M. Bender, S. Chakrabarti, and S. Muthukrishnan. Flow and stretch metrics for scheduling continuous job streams. In *Proceedings of the Ninth Annual ACM-SIAM Symposium on Discrete Algorithms*, pages 279–279, San Francisco, CA, January 1998.

[19] M. Bender et al. Unix for nomads - making unix support mobile computing. In *First Usenex Symposium on Location Dependent Computing*, August 1993.

[20] J. Bennett and H. Zhang. Wf2q: Worst-case fair weighted fair queueing. In *INFOCOMM'96*, March 1996.

[21] A. Bestavros. Aida-based real-time fault-tolerant broadcast disks. In *Proceedings of the IEEE Real-Time Technology and Applications Symposium*, Boston, MA, June 1996.

[22] A. Bhattacharya and S. Das. Lezi-update: An information-theoretic approach to track mobile users in pcs networks. In *Proceedings of the Fifth International Conference on Mobile Computing and Networking*, Seattle, Washington, August 1999.

[23] T.F. Bowen, G. Gopal, G. Herman, T. Hickey, K.C. Lee, W.H. Mansfield, J. Raitz, and A. Weinrib. The datacycle architecture. *Communications of the ACM*, 35(12), December 1992.

[24] R. Caceres and L. Iftode. Improving the performance of reliable transport protocols in mobile computing environments. In *IEEE Journal on Selected Areas in Communications*, pages 850–857, June 1995.

[25] J. Cai. *Information Retrieival in a Wireless Mobile Computing Environment*. PhD thesis, National University of Singapore, Singapore, 1999.

[26] J. Cai and K.L. Tan. Energy efficient selective cache invalidation. *ACM/Baltzer Wireless Networks*, 5(6):489–502, 1999.

[27] J. Cai and K.L. Tan. Tuning integrated disseminated-based information systems. *Data and Knowledge Engineering*, 30(1):1–21, 1999.

[28] J. Cai, K.L. Tan, and B.C Ooi. On incremental cache coherency schemes in mobile computing environment. In *Proceedings of the 13th International Conference on Data Engineering*, pages 114–123, April 1997.

[29] W.W. Chang and H.J. Schek. A signature access method for the starburst database system. In *Proceedings of the 15th VLDB COnference*, Amsterdam, The Netherlands, August 1989.

[30] M.S. Chen, P.S. Yu, and K.L. Wu. Indexed sequential data broadcasting in wireless mobile computing. In *Proceedings of the 17th IEEE International Conference on Distributed Computing Systems*, pages 124–131, May 1997.

[31] T. S. Chua, K. L. Tan, and B. C. Ooi. Fast signature-based color-spatial image retrieval. In *Proc. of the International Conference on Mutlimedia Computing and Systems '97*, pages 362–269, June 1997.

[32] H. Chung and H. Cho. Data caching with incremental update propagation in mobile computing environments. In *Proc. of the Australian Workshop on Mobile Computing and Databases and Applications*, Melbourne, Australia, February 1996.

[33] D. Comer. The ubiquitous b-tree. *ACM Computing Surveys*, 11(2):121–137, June 1979.

[34] A. Datta, A. Celik, J. Kim, D.E. VanderMeer, and V. Kumar. Adaptive broadcast protocols to support power conservant retrieval by mobile users. In *Proceedings of the 13th International Conference on Data Engineering*, pages 124–133, April 1997.

[35] A. Datta, D.E. VanderMeer, A. Celik, and V. Kumar. Broadcast protocols to support efficient retrieval from databases by mobile users. *ACM Transactions on Database Systems*, 24(1):1–79, March 1999.

[36] D. DeWitt, R. Katz, F. Olken, L. Shapiro, M. Stonebraker, and D. Wood. Implementation techniques for main memory database systems. In *Proceedings of the 1984 ACM-SIGMOD International Conference on Management of Data*, June 1984.

[37] C. Dhawan. *Mobile Computing: A Systems Integrator's Handbook.* McGraw-Hill, 1997.

[38] F. Douglis, R. Caceres, M. Kaashoek, P. Krishnan, K. Li, B. marsh, and J. Tauber. Storage alternatives for mobile computers. In *First Symposium on Operating Systems and Design*, November 1994.

[39] F. Douglis, P. Krishnan, and B. Marsh. Thwarting the power hungry disk. In *Proceedings of the 1994 Winter USENIX Conference*, 1994.

[40] H.D. Dykeman, M.H. Ammar, and J.W. Wong. Scheduling algorithms for videotex systems under broadcast delivery. In *Proceedings of the International Conference on Communications*, pages 1847–1851, 1986.

[41] P.K. Eng and K.L. Tan. Disseminating data in unreliable wireless environment. In *submitted for publication*, 1999.

[42] C. Faloutsos and S. Christodoulakis. Signature files: An access method for documents and its analytical performance evaluation. *ACM Transactions on Office Information Systems*, 2(4):267–288, 1984.

[43] J. Fernandex and K. Ramaritham. Adaptive dissemination of data in time-critical asymmetric communication environments. In *EuroMicro Conference on Real-Time Systems*, 1997.

[44] J. Flinn and M. Satyanarayanan. Powerscope: A tool for profiling the energy usage of mobile applications. In *Proceedings of the Second IEEE Workshop on Mobile Computing Systems and Applications*, February 1999.

[45] G.H. Forman and J. Zahorjan. The challenges of mobile computing. *IEEE Computer*, 27(6), April 1994.

[46] M. Franklin and S. Zdonik. Dissemination-based information systems. *IEEE Data Engineering*, 19(3):20–30, September 1996.

[47] M. Franklin and S. Zdonik. "data in your face": Push technology in perspective. In *Proceedings of the 1998 ACM-SIGMOD International Conference on Management of Data*, pages 516–519, June 1998.

[48] D.K. Gifford. Polychannel systems for mass digital communications. *Communications of the ACM*, 33(2), February 1990.

[49] A. Gurijala and U. Pooch. Propagating updates in asymmetric channels. In *Proceedings of the 1st International Workshop on Satellite-based Information Services*, November 1996.

[50] L. Hall. *Approximation Algorithms for NP-hard Problems*. PWS Publishing Company, 1997.

[51] S. Hameed and N. Vaidya. Log-time algorithms for scheduling single and multiple channel data broadcast. In *Proceedings of the 3rd ACM International Conference on Mobile Computing and Networking*, November 1997.

[52] S. Hameed and N. Vaidya. Efficient algorithms for scheduling data broadcastation. *ACM/Baltzer Wireless Networks*, 5(3):183–193, 1999.

[53] G. Herman, G. Gopal, K.C. Lee, and A. Weinrib. The datacycle architecture for very high throughput database systems. In *Proceedings of the 1987 ACM-SIGMOD International Conference on Management of Data*, pages 97–103, June 1987.

[54] Q. Hu, D. Lee, and W. Lee. Performance evaluation of a hierarchical data dissemination system. In *Proceedings of the 5th International Conference on Mobile Computing and Networking*, Seattle, USA, 1999.

[55] T. Imielinski and B.R. Badrinath. Querying in highly mobile distributed environment. In *Proceedings of the 18th International Conference on Very Large Data Bases*, pages 41–52, August 1992.

[56] T. Imielinski and B.R. Badrinath. Data management for mobile computing. *SIGMOD RECORD*, 22(1):34–39, March 1993.

[57] T. Imielinski and B.R. Badrinath. Mobile wireless computing: Challenges in data management. *Communications of the ACM*, 37(10), October 1994.

[58] T. Imielinski and H.F. Korth, editors. *Mobile Computing.* Kluwer Academic Publishers, January 1996.

[59] T. Imielinski and S. Viswanathan. Adaptive wireless information systems. In *Proceedings of the SIGDBS Conference*, pages 19–41, Tokyo, Japan, October 1994.

[60] T. Imielinski, S. Viswanathan, and B.R. Badrinath. Energy efficient indexing on air. In *Proceedings of the 1994 ACM-SIGMOD International Conference on Management of Data*, pages 25–36, June 1994.

[61] T. Imielinski, S. Viswanathan, and B.R. Badrinath. Power efficient filtering of data on air. In *Proceedings of the 4th International Conference on Extending Database Technology*, pages 245–258, March 1994.

[62] T. Imielinski, S. Viswanathan, and B.R. Badrinath. Data on air: Organization and access. *IEEE Transactions on Knowledge and Data Engineering*, 9(3):353–372, 1997.

[63] R. Jain and Y. B. Ling. A auxiliary user location strategy employing forwarding pointers to reduce network impacts of pcs. *Wireless Neworks*, 1:197–210, 1995.

[64] J. Jannink, D. Lam, N. Shivakumar, and J. Widom. Efficient and flexible location management techniques for wireless communication systems. In *Proceedings of the 2nd ACM International Conference on Mobile Computing and Networking*, 1996.

[65] S. Jiang and N. Vaidya. Scheduling data broadcast to "impatient" uers. In *International Workshop on Data Engineering for Wireless and Mobile Access*, Seattle, USA, August 1999.

[66] J Jing, A. Elmagarmid, A. Helal, and R. Alonso. Bit-sequences: An adaptive cache invalidation method in mobile client/server environments. *Mobile Networks and Applications*, 2(2):115–127, 1997.

[67] J. Jing, A. Helal, and A. Elmagarmid. Client-server computing in mobile environments. *ACM Computing Surveys*, 31(2):117–157, 1999.

[68] D. Johnson, A. Myles, and C. Perkins. Overview of mobileip. In *RFC*, April 1995.

[69] D. Karger, C. Stein, and J. Wein. *Handbook of Algorithms and Theory of Computation*. CRC Press, 1998.

[70] J. Kistler and M. Satyanarayanan. Disconnected operation in the coda file system. *ACM Transactions on Computer Systems*, 10(1), February 1992.

[71] D.E. Knuth. *The Art of Programming, Vol. 3: Sorting and Searching*. Addison Wesley, 1973.

[72] R. Kravets and P. Krishnan. Power management techniques for mobile communication. In *Proceedings of the Fourth International Conference on Mobile Computing and Networking*, pages 157–167, Dallas, TX, October 1998.

[73] G. Krishnamurthi, M. Azizoglu, and A. Somani. Optimal location management algorithms for mobile networks. In *Proceedings of the Fourth International Conference on Mobile Computing and Networking*, pages 223–232, Dallas, TX, October 1998.

[74] D.L. Lee, Y.M. Kim, and G. Patel. Efficient signature file methods for text retrieval. *IEEE Transactions on Knowledge and Data Engineering*, 7(3):423–435, 1995.

[75] W. Lee. *Mobile Cellular Telecommunication Systems*. McGraw-Hill, 1989.

[76] W.C. Lee, Q. Hu, and D. Lee. A study on channel allocation for data dissemination in mobile cpmputing environments. *ACM/Baltzer MONET*, 4(2):117–129, 1999.

[77] W.C. Lee and D. Lee. Using signature and caching techniques for information filtering in wireless and mobile environments. *Journal of Distributed and Parallel Databases*, 4(3):205–227, 1996.

[78] H.V. Leong and A. Si. Data broadcasting strategies over multiple unreliable wireless channels. In *Proceedings of the 4th International Conference on Information and Knowledge Management*, pages 96–104, November 1995.

[79] J.R. Lorch and A.J. Smith. Software strategies for portable computer energy management. *IEEE Personal Communications*, 5(3), June 1998.

[80] C. Perkins and P. Bhagawat. Routing over multi-hop wireless network of mobile computers. In *Proceedings of the ACM SIGCOMM'94*, 1994.

[81] E. Pitoura and P.K. Chrysanthis. Exploiting versions for handling updates in broadcast disks. In *Proceedings of the 25th VLDB COnference*, pages 114–125, Edinburgh, Scotland, September 1999.

[82] E. Pitoura and P.K. Chrysanthis. Scalable processing of read-only transactions in broadcast push. In *Proceedings of the 19th IEEE International Conference on Distributed Computing Systems*, pages 432–439, May 1999.

[83] E. Pitoura and G. Samaras. *Data Management for Mobile Computing*. Kluwer Academic Publishers, 1998.

[84] Pointcast Inc. WWW. Url:http://www.pointcast.com/. 1997.

[85] M.O. Rabin. Efficient dispersal of information for security, load balancing and fault tolerance. *Journal of the ACM*, 36(2):335–348, April 1989.

[86] S. Rajagoplan and B.R. Badrinath. An adaptive location management strategy for mobile ip. In *Proceedings of the 1st ACM International Conference on Mobile Computing and Networking*, October 1995.

[87] K. Ramamritham. Real-time databases. *International Journal of Distributed and Parallel Databases*, 1(2), 1993.

[88] N. Roussopoulous. An incremental access method for viewcache: Concept, algorithms, and cost analysis. *ACM Transactions on Database Systems*, 16(3):535–563, 1991.

[89] S. Salomone. Radio days. In *Byte. Special Issue on Mobile Computing*, 1995.

[90] R. Sandberg, D. Goldberg, S. Kleiman, D. Walsh, and B. Lyon. Designing and implementation of the sun network filesystem. In *Proceedings of the 1985 USENIX Summer Conference*, pages 119–130, June 1985.

[91] M. Satyanarayanan. Fundamental challenges in mobile computing. In *Proceedings of the 15th ACM Symposium on Principles of Distributed Computing*, Philadelphia, PA, May 1996.

[92] M. Satyanarayanan. Mobile information access. *IEEE Personal Communications*, 3(1), February 1996.

[93] M. Satyanarayanan, J.H. Howard, D.N. Nichols, R.N. Sidebotham, A.Z. Spector, and M.J. West. The itc distributed file system: Principles and

design. In *Proceedings of the 10th ACM Symposium on Operating Systems Principles*, pages 35–50, December 1985.

[94] M. Satyanarayanan, J.J. Kistler, P. Kumar, M. Okasaki, E.H. Siegel, and D.C. Steere. Coda: A highly available file system for distributed workstation environment. *IEEE Transactions on Computers*, 39(4), April 1990.

[95] J. Shanmugasundaram, A. Nithrakashyap, R. Sivasankaran, and K. Ramamritham. Efficient concurrency control for broadcast environments. In *Proceedings of the 1999 ACM-SIGMOD International Conference on Management of Data*, pages 85–96, June 1999.

[96] S. Sheng, A. Chandrasekaran, and R. E. Broderson. A portable multimedia terminal for personal communications. *IEEE Communications Magazine*, pages 64–75, December 1992.

[97] N. Shivakumar and J. Widom. User profile replication for faster location lookup in mobile environments. In *Proceedings of the 1st ACM International Conference on Mobile Computing and Networking*, pages 161–169, October 1995.

[98] M. Shreedhar and G. Varghese. Efficient fair queueing using deficit round robin. In *SIGCOMM'95*, Cambridge,MA,USA, 1995.

[99] A. Si and H.V. Leong. Query processing and optimization for broadcast database. In *Proceedings of the 7th International Conference on Database and Expert Systems Applications*, pages 899–914, September 1996.

[100] K. Stathatos, N. Roussopoulos, and J. Baras. Adaptive data broadcasting using air-cache. In *Proceedings of the 1st International Workshop on Satellite-based Information Services*, November 1996.

[101] K. Stathatos, N. Roussopoulos, and J. Baras. Adaptive data broadcast in hybrid networks. In *Proceedings of the 23rd International Conference on Very Large Data Bases*, August 1997.

[102] C. Su and L. Tassiulas. Joint broadcast scheduling and user's cache management for efficient information delivery. In *Proceedings of the 4th International Conference on Mobile Computing and Networking*, pages 33–42, Dallas, Texas, 1998.

[103] C.J. Su and L. Tassiulas. Broadcast schedule for information distribution. In *Proceedings of the IEEE INFOCOM'97*, pages 109–113, Kobe, Japan, April 1997.

[104] C. Tait and D. Duchamp. An efficient variable consistency replicated file service. In *Usenix File System Workshop*, pages 111–126, May 1992.

[105] K.L. Tan. Organization of broadcast data for effective and energy efficient join processing in wireless environments. *unpublished manuscript*, 2000.

[106] K.L. Tan. Organization of invalidation reports for energy-efficient cache invalidation in mobile environments. *ACM/Baltzer MONET (accepted for publication)*, 2000.

[107] K.L. Tan and J. Cai. Broadcast-based group invalidation: An energy efficient cache invalidation scheme. *Information Sciences*, 100(1-4):229–254, August 1997.

[108] K.L. Tan and J. Cai. An evaluation of cache invalidation strategies in wireless environments. *unpublished manuscript*, 2000.

[109] K.L. Tan and B.C. Ooi. Batch scheduling for demand-driven servers in wireless environments. *Information Sciences*, 109:281–298, 1998.

[110] K.L. Tan and B.C. Ooi. On selective tuning in unreliable wireless channels. *Data and Knowledge Engineering*, 28(2):209–231, 1998.

[111] K.L. Tan and J.X. Yu. Energy efficient filtering of nonuniform broadcast. In *Proceedings of the 16th IEEE International Conference on Distributed Computing Systems*, pages 520–527, May 1996.

[112] K.L. Tan and J.X. Yu. A break for workaholics: Energy-efficient selective tuning mechanisms for demand-driven-based wireless environment. In *Proceedings of the 5th International Symposium on Database Systems for Advanced Applications*, Melbourne, Australia, April 1997.

[113] K.L. Tan and J.X. Yu. A dynamic scheduler for the infinite air-cache. *Data and Knowledge Engineering*, 24(1):97–112, 1997.

[114] K.L. Tan and J.X. Yu. Generating broadcast programs that support range queries. *IEEE Transactions on Knowledge and Data Engineering*, 10(4):668–672, August 1997.

[115] K.L. Tan, J.X. Yu, and P.K. Eng. Supporting range queries in wireless environment with nonuniform broadcast. *Data and Knowledge Engineering*, 29(2):201–221, 1999.

[116] A.S. Tanenbaum. *Computer Networks*. Prentice-Hall, 1987.

[117] N. Vaidya and S. Hameed. Data broadcast scheduling: On-line and off-line algorithms. Technical Report 96-017, Computer Science Department, Texas A&M University, 1996.

[118] N. vaidya and S. Hameed. Data broadcast in asymmetric wireless environments. In *Proceedings of the Third International Conference on Mobile Computing and Networking*, 1997.

[119] N. Vaidya and S. Hameed. Scheduling data broadcast in asymmetric communication environments. *ACM/Baltzer Wireless Networks*, 5(3):171–182, 1999.

[120] M. Weiser, B. Welch, A. Demers, and S. Shenker. Scheduling for reduced cpu energy. In *First Symposium on Operating Systems and Design*, November 1994.

[121] J.W. Wong. Broadcast delivery. *Proceedings of the IEEE*, 76(12):1566–1577, December 1988.

[122] J.W. Wong and M.H. Ammar. Analysis of broadcast delivery in a videotex system. *IEEE Transactions on Communications*, 34(9):863–866, 1985.

[123] K.L. Wu, P.S. Yu, and M.S. Chen. Energy-efficient caching for wireless mobile computing. In *Proceedings of the 12th International Conference on Data Engineering*, pages 336–343, February 1996.

[124] P. Xuan, S. Sen, O. Gronzalez, J. Fernandex, and K. Ramaritham. Broadcast on demand: Efficient and timely dissemination of data in mobile environments. In *Proceedings of the the 3rd IEEE Symposium on Real-Time Technology and Application*, Montreal,Canada, 1997.

[125] J. Yu, T. Sakata, and K.L. Tan. Statistical estimation of access frequencies in data broadcasting environments. *ACM/Baltzer Wireless Networks (accepted for publication)*, 2000.

[126] J.X. Yu and K.L. Tan. An analysis of selective tuning schemes for nonuniform broadcast. *Data and Knowledge Engineering*, 22(3):319–344, May 1997.

Index

$(1,m)$ index, 46

access pattern, 121
nonuniform, 121
uniform, 121
access time, 1, 19, 23, 36, 45, 73, 149
active mode, 43, 150
adaptive, 67
algorithm CF, 56
asymmetric communication, 4
asymmetric communication environments, 117
asynchronous, 90

batch, 13
batching, 67
battery, 4, 46, 89
bcast wait, 45
broadcast channel, 43
broadcast cycle, 24, 140
copy, 24
frequency, 24
spacing, 24
broadcast disks, 30
broadcast epoch, 129
broadcast organization
fault-tolerant, 73–88
broadcast program, 24, 117, 149
flat, 24
generation, 26–35
non-flat, 25–29
offline algorithm, 29
online algorithm, 28
packet fair queuing, 29
priority index policy, 29
Vaidya and Hameed's algorithm, 28–29
Wong's algorithm, 26–28
non-flat coalesced, 29–32
broadcast disks, 30
bucketing algorithm, 30
Vaidya and Hameed's algorithm, 30–31
properties, 24–26
range queries, 32–35
algorithm NrBP, 33
regular, 25
skewed, 25
broadcast window, 93
bucket, 44

cache, 89
cache coherency
relational operation, 133–139
cache invalidation, 89–116
asynchronous, 90
bit-sequences, 105–107
bit-sequences with bit count, 108–111
broadcast timestamp, 97
dual-report invalidation, 101–105
GCORE, 99–101
query-level, 108
selective dual-report invalidation, 107–108
stateful, 89
stateless, 90
synchronous, 90
taxonomy, 90–97
cache-level invalidation, 93
canned queries, 134
chronic starvation, 130
client view, 134
constant broadcast size, 130
constant fanout, 56
contention, 89
corrupt, 73
cues, 63, 65
cyclic redundancy code, 73

Data delivery mechanisms
bandwidth-oriented, 14–15

dynamic allocation, 15
static allocation, 15
broadcast disks, 16
data-oriented, 10–12
demand-driven, 11
publication, 10
email listing, 16
email listing digest, 16
integrated methods, 20
mechanics-oriented, 12–13
broadcast, 13
client-initiation, 12
event-driven, 13
schedulability, 13
schedule-driven, 13
server-initiation, 12
unicast, 13
organization-oriented, 13–14
broadcast program, 13
selective tuning, 14
polling, 17
publish/subscribe, 16
pull-based methods, 18
push-based methods, 17
request/response, 16
taxonomy, 9–19
data delivery mechanisms, 1
database server, 2
demand-driven, 62
demand-driven group, 121
differential file, 135
direct offset, 58
directory, 44
disconnection, 89
disk-based partition join, 141
displacement, 82
dissemination-based information system, 117
distributed index, 59
distributed indexing, 47
doze mode, 43, 150
dumb terminals, 2
dynamic, 11

energy, 89
energy consumption, 1
energy units, 113
error-prone, 73

failure probability, 75
false drop, 68
false invalidation, 92
fault, 73
congested network, 74
electromagnetic noise, 73
fault-tolerant, 73
FIFO policy, 64
final probe, 74
flexible index, 53, 85
control index, 54
global index, 54, 85
frigid state, 128
full index, 58

gap, 50
group invalidation report, 103
group-based algorithm, 68

hash bucket, 50, 82
hash function, 50
hash-based index, 49
hashing B protocol, 50
history, 92

ignore factor, 130
imbalanced tree index, 56
incremental, 134
incremental log, 135
index, 44, 93
index replication, 75
indirect offset, 58
Information dissemination, 1
models, 9–21
performance metrics, 19–21
access time, 19
packet efficiency, 19
power efficiency, 19
responsiveness, 19
robustness, 19
tuning time, 19
initial probe, 74
integrated model, 117–131, 150
adaptive, 121–131
architecture, 118
dynamic, 121–131
static, 118
integrated signature, 52
inter-index scheme, 76
all-children replication, 76
hash-based, 82
self replication, 78
intra-index scheme
distributed index, 79
tree-based, 79
intra-index schemes, 78
invalidation report, 90
compressed, 93
content, 92–93
granularity, 92–93
organization, 93
selective tuning, 93–94
size, 93
update history, 93
invalidation mechanism, 93–94
participants, 94
scale, 94

log structure, 94–95
log history, 95
signature, 93
IP-address, 64, 66
IP-based algorithm, 66

join, 139–146
greedy algorithm, 143
nested-loops join, 140
partition-based join, 141

lifespan, 89
liquid warm state, 128

marginal gains, 128
materialized view, 134
mobile, 89
mobile computing, 1
mobile support station, 2
multi-level signature, 52
multicasting, 67

noise, 73
non-flat coalesced, 25
nonreplicated roots, 47

object invalidation report, 98, 103
offset, 44

partition, 49
partitioned attribute, 141
performance metrics, 36
periodic push, 9
piggyback, 126
full piggybacking, 126
partial piggybacking, 126
popularity factor, 130
portion, 60, 61
power, 4
preemptive, 40
priority, 130
priority index policy, 131
probe time, 45
program, 23
publication group, 121
pull channel, 118
pull mode, 23, 117
pull model, 35–41
pure pull, 118
push channel, 118
push mode, 23, 117
push model, 23–35
push-based system, 139

query arrival time, 113
query-level invalidation, 93

relational operation, 133–146
join, 133
reprobe, 74
request-response, 9
residence latency, 130
responsiveness, 36
restart, 74
robustness, 37

scale, 36
schedule, 23, 63
scheduling algorithm, 23
scheduling overhead, 37
scheduling strategy, 35, 149
RxW, 38
earliest deadline first, 40
earliest first request, 37
longest total stretch first, 40
longest wait first, 38
MAX, 40
most request first, 37
preemptive longest wait first, 40
preemptive remaining service time, 40
priority index policy, 39
shortest service time first, 39
section, 54, 60, 61
Security, 4
segment, 53, 60, 85
segment-level index, 58
selective tuning, 14, 43–69, 73, 93–94, 108, 139, 150
pull-based, 62–69
push-based, 43–62
stateful scheme, 63–65
stateless scheme, 55, 65
serializable, 151
shift pointer, 50, 82
signal distortion, 73
signature-based index, 51
simple signature, 52
stable environment, 12
stateful server, 63, 89
stateless server, 63, 90
static, 11
stretch, 36
superimpose, 51
synchronous, 90

temperature, 128
tree-based index, 47
tuning time, 19, 45, 73

unpartitioned attribute, 144
unreliable, 150
update arrival time, 115
update invalidation, 92, 135
update log window, 95
update propagation, 92, 135

uplink channel, 19

vapor state, 128
variable broadcast size, 130
version, 44
view, 134
view invalidation, 134
view reconstruction, 134

walkstations, 2
wired network, 2
wireless cell, 2
wireless channel, 3
    downlink channel, 3
    uplink channel, 3
wireless computing, 1
    applications, 5–7
        finance, 5
        health care, 5
        law enforcement, 6
        manufacturing, 5
        transport, 6
    benefits, 5–7
    characteristics, 3–5
    design issues, 7–8
    potential, 5–7
    reference architecture, 2–3
wireless networks, 1

About the Authors

Kian-Lee Tan, received his B.S., M.S. and Ph.D. in computer science, from the National University of Singapore, in 1988, 1991 and 1994 respectively. He is currently an Associate Professor in the Department of Computer Science, National University of Singapore. His major research interests include multimedia information retrieval, wireless computing, query processing and optimization in multiprocessor and distributed systems, and database performance. He has published over 60 papers in international conferences and journals. He is a co-author of the tutorial (IEEE CS Press, 1994) entitled "Query Processing in Parallel Relational Database Systems", and the monograph (Kluwer Academic Publishers, 1997) entitled "Indexing Techniques for Advanced Database Systems".

Beng Chin Ooi received his B.Sc. and Ph.D in computer science from Monash University, Australia, in 1985 and 1989 respectively. He was with the Institute of Systems Science, Singapore, from 1989 to 1991 before joining the Department of Computer Science at the National University of Singapore, Singapore. His research interests include database performance issues, database UI, multi-media databases and applications, and GIS. He is the author of a monograph "Efficient Query Processing in Geographic Information Systems" (Springer-Verlag, 1990), and co-author of the tutorial "Query Processing in Parallel Relational Database Systems" (IEEE CS Press, 1994) and monograph "Indexing Techniques for Advanced Database Systems" (Kluwer Academic Publishers, 1997). He has published many conference and journal papers and serves as a PC member in a number of international conferences. He is currently on the editorial board of the following scientific journals: International Journal of Geographical Information Systems, Journal on Universal Computer Science, Geoinformatica and International Journal of Information Technology.

MIX
Papier aus verantwortungsvollen Quellen
Paper from responsible sources
FSC® C105338

If you have any concerns about our products,
you can contact us on
ProductSafety@springernature.com

In case Publisher is established outside the EU,
the EU authorized representative is:
Springer Nature Customer Service Center GmbH
Europaplatz 3, 69115 Heidelberg, Germany

Printed by Libri Plureos GmbH
in Hamburg, Germany